Praise for
THE END OF PLENTY

"Joel Bourne, who grew up working on his family's farm, traveled the world to explore what may be the greatest challenge facing the next generation. The result is calm, lucid—and fascinating."

—CHARLES C. MANN,
author of *1491* and *1493*

"Here is a wake-up call, and also a call to action. The stakes could not be higher: To stave off apocalypse, we must grow a whole lot smarter in a hurry—starting by heeding the cutting-edge wisdom contained in Joel Bourne's richly researched and passionately argued report from the Malthusian margins."

—HAMPTON SIDES,
editor-at-large for *Outside* magazine
and author of *In the Kingdom of Ice*

"A thoroughly researched and exceptionally thoughtful and balanced look at the consequences of industrial farming. [Bourne's] book should convince every reader of the compelling need to address world food problems through more skillful and sustainable agronomy, but also through education, especially of women, and universal family planning."   —MARION NESTLE,
professor of nutrition, food studies, and public health,
New York University, and author of *Food Politics*

"An important read for everyone."   —PAUL R. EHRLICH,
coauthor of *The Dominant Animal*

"In a well-documented and fast-moving manner, Joel Bourne Jr., one of America's foremost experts by virtue of his 'hands-on' experience, education and world travel, clearly depicts a strategic challenge for America's national security in the coming years."

—HENRY H. SHELTON,
General, US Army (retired),
14th Chairman Joint Chiefs of Staff

"This book was a stark reminder that Moore's Law does not apply to food production technology and that there is a very real risk we may be headed towards a future in which digital abundance coexists with a scarcity of tangible resources that are essential for human survival and wellbeing."   —MARTIN FORD,
Financial Times, Best Books of 2015

"As in a good *National Geographic* story, Bourne mixes telling anecdotes with facts and figures to paint a picture of how the world's food supply is changing."   —BEN STEELMAN,
StarNews

"Bourne is most persuasive when he brings his training in agronomy to serve up surprising insights into the future of food."

—RAJ PATEL,
New York Times Book Review

"Joel K. Bourne brings something refreshing to the interminable debate about how we are going to feed a human population that is expected to reach 9 billion in the next few decades—objectivity."

—BARRY ESTABROOK,
Orion

THE END OF PLENTY

THE RACE TO FEED A CROWDED WORLD

Joel K. Bourne Jr.

W. W. Norton & Company
Independent Publishers Since 1923
New York • London

For information about permission to reproduce selections from this book,
write to Permissions, W. W. Norton & Company, Inc.,
500 Fifth Avenue, New York, NY 10110

For information about special discounts for bulk purchases, please contact
W. W. Norton Special Sales at specialsales@wwnorton.com or 800-233-4830

Manufacturing by RR Donnelley, Harrisonburg
Book design by Chris Welch
Production manager: Louise Mattarelliano

Library of Congress Cataloging-in-Publication Data

Bourne, Joel K., Jr.
The end of plenty : the race to feed a crowded world / Joel K. Bourne Jr. — First ediiton.
    pages cm
Includes bibliographical references and index.
ISBN 978-0-393-07953-1 (hardcover)
1. Food supply—Forecasting. 2. Food consumption forecasting. 3. Food security.
I. Title.
HD9000.5.B58 2015
363.8—dc23
                                                    2015001552

ISBN 978-0-393-35296-2 pbk.

W. W. Norton & Company, Inc., 500 Fifth Avenue, New York, N.Y. 10110
www.wwnorton.com

W. W. Norton & Company Ltd., Castle House, 75/76 Wells Street, London W1T 3QT

1 2 3 4 5 6 7 8 9 0

For my mother, who taught me to love language.

For my father, who taught me to love the land.

And for all those who toil each day to feed the world.

CONTENTS

PART I

PART II

PART I

Egyptians mob a kiosk selling government-subsidized bread near the Great Pyramid of Giza as food prices soar in 2008.

The Erstwhile Agronomist

People are fighting. Killing for bread, some are even pulling out knives. What is happening? What is this? Famine?
—*Egyptian man in breadline, March 2008*

E very now and then someone asks me what I studied in college. Since I've spent most of my career as a writer for *National Geographic*, they expect to hear English or journalism, or perhaps something even more esoteric. I reply, "Agronomy," and typically get blank stares. When I explain that it's a combination of soil science and plant science for row crop production, their interest wilts. Journalists call this MEGO—short for "my eyes glaze over." We apply the term to topics that are dull, complex, and boring. Journalism, for whatever reason, is fascinating. Agronomy, alas, is MEGO.

I've come to believe that few people in my home country know, much less care, what an agronomist is these days. Those that do tend to view the discipline with fear or loathing—the realm of migrant workers, chemical agriculture, and "Big Food." It doesn't matter that nearly every morsel they put in their mouths—every crumb that gives them life and vigor—was likely produced through the art and science of agronomy. The first five presidents of the United States were all farmers, as obsessed with new seeds and breeds as we are with the latest cell phone. Thomas Jefferson even risked his life to smuggle rice seeds out of Italy—an offense punishable by death at the time.

Today the cities most of us call home would be concrete waste-lands if it weren't for a steady stream of victuals pouring in from the farms and fields that now blanket 40 percent of the Earth's dry land. I once covered a New York City mayoral election from Hunts Point in the South Bronx, at the time one of the most blighted urban neighborhoods in the nation. I'll never forget the empty crack vials in the gutters, or the endless line of semis threading their way down the streets. Each day 15,000 trucks stream in and out of the Hunts Point Terminal Market, delivering produce from 49 states and 55 countries to keep New Yorkers fed. Imagine the chaos that would ensue if those trucks ever stopped coming. Though I love my profession and believe it is fundamental to a functioning democracy, a nation can survive without journalists. It cannot last a day without farmers.

The world contains more than 50,000 edible plants, but only three—wheat, rice, and corn—directly or indirectly (through livestock feed) provide 80–90 percent of all the calories that humans consume. For most of human history, food demand was pretty steady, driven by world population growth. The more mouths to feed and the more farmers there were (since most occupations before the industrial revolution were agricultural), the more land

farmers cleared to grow crops to feed them. Though religious freedom was allegedly the big draw for colonists to the United States,
the tantalizing pull of free farmland helped fill the ships with thousands of landless peasants willing to risk the hardships of colonial
life for a few acres of their own.

For the most part, farmers have more than kept up. Even during
the staggering population boom of the last half century, when the
number of people on the planet skyrocketed from 3 billion in 1960
to 6 billion in 2000—the fastest doubling in human history—our
annual grain production rose even faster, nearly tripling during
the same period. Unlike previous centuries, when more food was
grown by cutting more forests and plowing more plains to create
more farmland, this time the increase came mostly from steadily
increasing yields on lands already in production. Better seeds, combined with more fertilizer, pesticides, and irrigation, enabled farmers to grow more crop from each acre. Farmers grew so much extra
food during the 1960s that they actually helped alleviate global
poverty by making food cheaper in most places around the world.
The change was so dramatic it was dubbed the "green revolution."

Despite the enormous impact that agriculture has had on our
lives and the planet, many urbanites and suburbanites in the
United States look upon the rural hinterlands, and the 2 percent of
the population that now tends them, with a certain amount of disdain. When my father was a boy in the 1920s, nearly a third of the
US population lived on farms. Today, the closest most people get
to country is when they hit the wrong button on the car radio dial.
And this is not just a US phenomenon. In a 2013 survey of more
than 27,000 primary-school children in the United Kingdom, one
in three thought cheese was a vegetable, and one in five thought
pasta came from animals. Another survey found that a quarter of
Australian sixth-graders thought yogurt grew on trees.

Our estrangement from the land isn't unusual. In fact, our

increasing urbanization is viewed in economic circles as downright beneficial. One of the main economic criteria used by the World Bank to distinguish rich developed nations from poor developing ones is how *few* people actually work in agriculture. In the eyes of economists, the fewer farmers the better.

I understand why they do this, but it still rubs me wrong. For nearly half my life a farmer was all I aspired to be. I grew up in a rural town in eastern North Carolina and spent much of my youth working and playing on my grandfather's farm in the county. I hunted quail, squirrels, and wood ducks in the pines, hardwood sloughs, and swamps, and fished for bass and bream in the ponds and bordering river. I worked in tobacco with the children of the black families that had lived on the farm since it was a plantation. I chopped weeds out of the cotton, mowed pastures, built fences, planted wildlife plots, helped my father harvest honey from his hives, and picked up trash along the highway. I learned how to drive a tractor years before I learned to drive a car. I don't remember ever deciding to be a farmer. I just assumed it was my destiny. I applied to only one college, North Carolina State University, and enrolled in the School of Agriculture and Life Sciences in the fall of 1981.

I landed on the brick-covered campus like Brer Rabbit thrown in the briar patch. I went to class in work boots, jeans, and a baseball cap with a patch that read, "Pride in Tobacco." I'd get up early to work half a day on the university test farm picking apples, mowing fields, cleaning manure out of chicken houses, and typically fall fast asleep in chemistry class after lunch. I drove an old four-wheel-drive Scout—a jeep-like vehicle made by the tractor company International Harvester—festooned with canoe racks and an assortment of shotgun shells and fishing lures rattling around the glove box. One summer I worked for the Department of Soil Science measuring soil moisture in cornfields across the state; the next, for the Department of Entomology trapping aphids that spread tobacco mosaic virus in the state's biggest money crop. I even played in a bluegrass band that a friend, after more than a few swigs of moonshine, had nicknamed "Boy Joe and the Agriculture Club."

The more I learned about production agriculture, however, the less certain I became. As a boy I'd watched my father clean up the family farm after my grandfather passed away, taking out the old hedgerows that provided such great habitat for the quail I loved to hunt. He sold my grandfather's herd of Hereford and Angus cows, whose hungry lowing had been the baritone chorus of my youth, and plowed their pastures into fields. This was 1972, when Secretary of Agriculture Earl Butz told every farmer in America to plant fencerow to fencerow. Farming was no longer a family affair. It was *agribusiness*. The American farmer was going to feed the world.

Over the years I'd watched as that rich, black bottomland soil turned into light-gray sand—little more than a growth medium to hold the crop roots that were fed by heavy applications of fertilizer. The river that had once flowed deep and clear around the pastures became a chocolate milkshake filled with eroded soil every time it rained. After the eradication of the boll weevil, cotton grew more plentiful. But it still needed frequent dousing with insecticides. The number of coveys I could find with my half-trained bird dogs dropped from four or five in a day, down to two or three, and then finally one, if we were lucky. When I asked a wildlife expert from the university what was going on, he told me that planting fencerow to fencerow had destroyed the habitat the quail loved, while the pesticides had poisoned the birds outright, weakened them so they forgot their defensive instincts, or so reduced the insect populations that baby quail chicks had little to eat. Quail, tough little birds that had flourished on southern farms since colonial days, were no match for agribusiness.

My college adviser was a barrel-chested agronomist from Pennsylvania named Bill Fike, a man I greatly respected. On a whim one day, I asked him about organic agriculture, which generated a good belly laugh from his substantial belly. "Fine for your garden," he said, "but it'll never feed the world." My classmates were mostly farm boys who had been driving sprayers almost as long as they'd been riding bikes. Many of them worked for agricultural chemical companies in the summer selling farmers the latest weed or insect killers. They all firmly believed that they were providing a much-needed service—giving good, scientific advice to less educated farmers. It was part of the escalating chemical war waged on weeds and bugs that each year destroyed a third of our food supplies. What could be wrong with that?

I remember one day sitting in a weed science class during my junior year. It was a warm October afternoon and the professor had cracked a few windows to let in the sweet Carolina air. He was one of the senior faculty, famous among the student Agronomy Club for using the same mimeographed tests from the previous decade and going off on tangents when the mood struck. That afternoon, someone got him going with a question about DDT.

"The greatest chemical ever invented," he said with surprising vehemence. "It saved millions of lives during World War II and afterwards against malaria. Rachel Carson was just a goddamn kook . . . part of a multi-million-dollar PR campaign to smear a perfectly good chemical. There was no scientific basis for that book at all."

I'd never read Rachel Carson, but I knew the gist of *Silent Spring*. My father loved to fish almost as much as he loved working on the farm. We'd spent a lot of time bobbing around the North Carolina coast in our old wooden skiff from the late 1960s onward. I remembered the wonder I felt when I spotted my first brown pelican skimming over the waves like a pterodactyl from my dinosaur book. It was soon followed by the great fish hawks the ospreys, and even bald eagles that once again began nesting in the tops of tall cypress trees along Pamlico Sound—birds even my father hadn't seen since his youth in the 1930s. I thought it was common knowledge that DDT had accumulated in their food chain, causing thinning eggshells and high chick mortality, and that they were now rebounding a decade after the chemical had been banned. I cautiously raised my hand from the back of the room.

"If that's the case, why are we just now seeing pelicans, ospreys, and eagles coming back to the coast?"

The professor turned on me as if I'd just spat in his coffee.

"Wildlife?" he shouted with complete disdain. "It's all cyclical!" That was it. End of discussion.

I was shocked. I knew that he'd spent much of his career testing the effectiveness of the latest pesticides, but this putative man of science was wearing chemical blinders. For the first time I realized that even tenured professors at large land-grant universities could be full of crap—and that I was just a few semesters away from a degree that I no longer believed in.

EARL BUTZ'S CALL to agricultural arms a decade earlier exacerbated my dilemma. The great plow-up had led to massive grain surpluses, rock-bottom commodity prices, and so many farmers going bankrupt that by 1985, Willie Nelson, John Mellencamp, and Neil Young staged their first Farm Aid concert to help them out. Though my father had worked some of the land in the 1950s, his father soon forced him to go to law school instead. We had farmed on shares with tenant farmers ever since. The tractor and planter we owned for mowing paths and sowing wildlife plots were too small for full-time use, not to mention being old and worn-out.

I did the math. If I was serious about farming, I would have to go to the bank and borrow about $250,000 just to buy the equipment I would need to start. My father would have to sign the note and put up the farm for collateral. I'd have to pay cash rent to my dad and uncle to replace what they were getting from the tenant farmers. I was staring at a mountain of debt just to get into the business when some of the best farmers in the county were going bankrupt all around me. And if I failed, we could lose the farm.

I wasn't alone in my economic predicament. Only a few of my classmates actually returned home to farm; most found better, less stressful, and higher-paying jobs working as pesticide sales reps for agricultural chemical companies. Others became county agents or crop consultants, which amounted to about the same thing.

I didn't like any of these options. So after graduation I hit the road. I sold my Scout, bought a plane ticket to Hawaii, and helped build hog fences in Hawai'i Volcanoes National Park. I then joined a crew of mostly British kids on an old tall ship named *Zebu*, after the African cow, that we sailed to Australia. I left the group in Sydney, bought a $500 car, and drove up to the Queensland outback, where I found work as a tractor driver on a big sheep and wheat station. I helped with the shearing and disked fields so big I'd make only two circuits before lunch and another two before dark.

I made enough money in the outback to get to New Zealand's North Island, where I hitchhiked down to the farming community of Te Awamutu. There I volunteered to work on a sheep and dairy farm owned by a young couple who had five kids and were struggling to make ends meet. Milk and sheep prices were so low that Mike, a young farmer with rugged Kiwi good looks, had stooped to modeling for a clothing catalog, which spurred no end of ribbing at the local pub. I spent a few weeks helping him vaccinate sheep and milk his dairy herd amid rural valleys more beautiful than any I'd seen before, with deep-green hills covered in clover and sprinkled with white sheep.

After New Zealand I traveled to Indonesia and walked through the terraced rice fields where women and children, their backs bent, ankle deep in water, traipsed by in endless servitude to the crop. On every farm I visited, I saw the same toil, the same struggle, the same anxiety, whether it was drought or poor prices or mounting debt. By the time I got back home, I still loved the farm, but I wanted no part of farming. So I did what millions of other kids had done before me. I got a job in the city and never looked back.

I couldn't completely shed my roots. One of the first jobs I landed in journalism was working for agricultural trade magazines. I'll never forget my first day at Columbia University's Graduate School of Journalism, when every student was asked about their prior journalism experience. Several had worked as interns at the *New York Times* or *Washington Post* or major news networks. Some were midcareer journalists back for their master's. When my turn came, the blood rushed to my face. "I was most recently the field editor of *Rice Journal*, *Peanut Farmer*, and *Flue-Cured Tobacco Farmer* magazines," I drawled, to the great amusement of my classmates. I even wrote my master's thesis on the Long Island Sound lobster industry, spending freezing days on the water pulling pots with my subjects.

In retrospect, my background served me well. At *National Geographic* I was frequently tapped to cover rural issues—whether Louisiana shrimpers suffering from wetland loss, Central Valley farmers struggling through record drought, or sugar barons producing biofuel in Brazil. I told my colleagues it was because I was the only one on staff who spoke fluent "redneck." Growing up around people who worked hard to wrest a modest living from the land and waters gave me enormous sympathy and respect for their plight. I never pulled punches when covering their industries. Instead, my sympathies for the working men and women made me dig even harder into the corporate and government policies at the top that were putting those at the bottom—as well as you and me—at greater risk.

Like most journalists, I was covering random environmental brush fires around the world. It wasn't until I was assigned to cover the global food crisis in 2008 that all the disparate issues I'd reported on over the years came into clear focus. After decades of producing surpluses and low food prices in many nations, the green revolution was over, leaving unsustainable monocultures and ecological destruction in its wake. More ominously, the system of agriculture I'd been trained in was no longer capable of feeding the 7 billion people on the planet. Pandemonium ensued. Agronomy was MEGO no more.

THAT SPRING, TENS of thousands of Cairo residents walked out into the streets before dawn and performed a painful ritual that would have been familiar to the pyramid's builders nearly 5,000 years before. They formed long lines outside of government bakeries and fought each other to buy bread. Several people died in the riots.

Egyptians eat more wheat per capita than anyone else in the world. Bread is so important there it goes by the name of *aish*, also the word for "life." In the six months leading up to the riots, the world market price of wheat soared to $13 a bushel, the highest in inflation-adjusted dollars since World War I. The price of bread from Egypt's private bakeries rose fivefold. People could still buy subsidized *baladi*, the flat, round loaves known as "country bread," for a fixed price of five piastres, or one US cent, far below the cost of the wheat within it—if they could endure the long lines to get it.

During the 1950s and early 1960s, Egypt's farmers grew enough wheat to feed the nation and even export some to their neighbors. No more. Although the high dam at Aswan, finished in 1970, expanded their irrigated area and helped triple wheat yields between the 1950s and 1990s, the number of Egyptians more than quadrupled, from 20 million in 1950 to 83 million in 2010. Farmers could not keep up. Today they harvest little more than half of the 18 million tons that Egyptians eat each year, forcing the nation to import the rest. Egypt is now the world's largest importer of wheat.

The Egyptians were not alone. Between 2005 and 2008, the international prices of wheat and corn—known as "maize" outside the United States—tripled, while the price of rice shot up fivefold. Protesters in Haiti flooded the streets soon after the Egyptians rioted, breaking down the gates of the presidential palace. Five died, including a UN peacekeeper, all because the price of rice, beans, and cooking oil had soared out of reach. The rioters ultimately forced the resignation of Prime Minister Jacques Edouard Alexis, who was powerless to reduce food prices. It was the same story in Cameroon, where 40 people were killed. The same in Bangladesh, in Bolivia, in Burkina Faso and Ivory Coast. The same in Mexico, Yemen, Pakistan, Uzbekistan, Sri Lanka, Senegal, and Somalia.

Violent protests occurred in more than a dozen countries around the world, while many other countries took drastic action to prevent similar riots at home. Russia pressured retailers into freezing the price of basic commodities before the 2007 elections to quell growing anger at rising food prices. Vietnam, one of the world's largest exporters of rice, banned all rice exports to preserve domestic stocks, as did India and Thailand.

Even in the United States, nearly 15 percent of households found themselves "food insecure"—the highest percentage since the US Department of Agriculture had begun measuring the statistic in 1995. The number of US citizens receiving food stamps had hovered between 20 and 25 million since 1980, but as unemployment rose during the financial crisis in 2008, more and more people needed help to feed their families. By 2012 the number had doubled, to more than 46 million.

Thanks to the global grain trade—which has been going on at least since Egypt was the bread basket for the Roman Empire—price spikes are usually short-lived, evened out by the flow of grain from countries with surpluses to those with shortfalls. Only a handful of countries, however, are fortunate enough to produce surplus grains. Bad weather in Brazil, the United States, or Russia can cause shortages that make prices rise, but they quickly fall as governments import more grain from elsewhere and farmers respond to higher prices by planting more of the crop. It's classic supply and demand. The conventional wisdom has long been that the best cure for high grain prices is high grain prices.

This time was different. In 2007 and 2008 the world's farmers reaped near-record grain harvests. Something was fundamentally wrong. "This is not a supply shock this time," said Chris Barrett, an agricultural economist at Cornell University in the midst of the crisis. "This is being driven by a long-term imbalance between production and consumption."

In other words, the world is running out of food.

IN 1992, AN estimated 824 million people were considered malnourished around the world. The problem was so dire that at the United Nations World Food Summit of 1996, the developed countries committed to halving world hunger by 2015.

Progress has been pitifully slow. A month from that mark, the number of hungry has fallen by less than 20 million. If the 805 million malnourished people on the planet were gathered into one nation, it would have the third largest population, behind China and India—almost the size of the populations of the United States and the European Union combined. As horrific as that is, more than twice that number of people are undernourished, unable to get essential vitamins and minerals from the foods they eat. In 2014, some 2 billion people, mostly poor women and children, suffered such "hidden hunger," often leading to stunting, blindness, cognitive problems, susceptibility to disease, and early death.

Poverty is partly to blame. The world's farmers produce enough calories today to feed 9 billion people a healthy, 2,700-calorie-per-day, mostly vegetarian diet. Unfortunately, the bulk of those calories are grown on fertile, well-watered lands half a world away from the people who need them most, and they cost money to transport. When nearly half the planet lives on less than two dollars per day—like 16 million Egyptians—price spikes can make three, or even two meals a day impossible to afford.

The food riots in Haiti in 2008 revealed the pitfalls of relying on free trade to provide a poor nation with its staple grain. Haiti had been self-sufficient in rice as late as the mid-1980s. But in 1994, when President Bill Clinton restored ousted President Jean Bertrand Aristide to power, he requested that Aristide drop Haiti's protective tariffs on imported rice. The country was soon flooded with cheap "Miami rice" from the United States, a crop that is heavily subsidized and grown in just a few states, including Clinton's home state of Arkansas. The imports destroyed local rice production—Haiti's small farmers simply couldn't compete—and left the poorest nation in the Western Hemisphere at the mercy of the international grain market. From then on, Haiti imported the bulk of its rice from the United States, while the US Agency for International Development (USAID) tried to help Haiti's farmers grow cash crops like coffee and mangoes to sell back to American shoppers. US diplomats defended the policy even as Haitians were rioting in the streets and filling their children's stomachs with cookies made from butter, salt, and dirt.

Two years after the riots, former president Clinton—now a UN special envoy to Haiti—publicly apologized for the rice policy that had done so much harm to Haiti's people, as well as for the impact of decades of agricultural trade liberalization on poor countries around the globe.

"It has not worked," Clinton said bluntly. "It may have been good for some of my farmers in Arkansas, but it has not worked. It was a mistake. . . . I have to live every day with the consequences of the lost capacity to produce a rice crop in Haiti to feed those people, because of what I did. Nobody else. . . . And it's failed everywhere it's been tried. You just can't take the food chain out of production."

The other, more worrisome problem is global grain reserves. Most years the world's farmers easily grow more grain than the world's consumers can use. The surplus is stored in grain elevators or warehouses as insurance against a bad crop the following year. These reserves also help keep prices down. Cereal grain production fell short of consumption only three years during the 1960s, and three years during the 1970s. The number of deficit years rose to five during the dry 1980s and fell back to four years in the 1990s.

Since 2000, however, the world has consumed more grain than it has grown in 8 years out of 12, whittling down global stockpiles to less than 70 days of consumption—the lowest levels since the mid-1970s. In 2007, world grain reserves fell to a 61-day supply, the second lowest level on record, helping fuel the panic that drove the food price crisis. The reasons for the drawdown are multiple and complex, but the trend is clear. Demand is slowly outstripping supply.

Population growth is the classic driver of food demand. The

world has nearly 80 million more mouths to feed each year, and we will have another 2.4 billion people by midcentury. That's like adding another China and India to the global table. While poverty is part of the problem, it's compounded by the fact that much of the world is getting richer. When people have more money, they tend to eat more meat and dairy products. It takes five times more grain to get the equivalent amount of calories from pork as it does from simply eating the grain itself. Ten times more in the case of grain-fattened beef. More than two-thirds of the world's agricultural land is already used to grow feed for livestock, yet world meat consumption is on track to *double* by 2020. If everyone in the world ate as much meat as Americans do (176 pounds per person per year), we'd need to find another planet to raise the feed and fodder for all our livestock.

We are not simply feeding more food to ourselves and our farm animals. We are now feeding it to our cars as well. Ethanol distilleries currently consume more than a third of the US corn crop; land devoted to biofuel crops is projected to increase fourfold by 2030, rising to 10 percent of all arable land in the United States and 15 percent of farmland in Europe.

With grain consumption spurred by increases in livestock, cars, and people, the UN's Food and Agriculture Organization projects that we'll need to boost grain production by at least 70 percent by 2050. Other agricultural experts put the number even higher, at 80–110 percent. Gebisa Ejeta, a plant breeder at Purdue University who won the 2009 World Food Prize—the Nobel for aggies— put that number into perspective: "We'll have to learn to produce as much food in the next four decades as we have since the beginning of civilization."

That in itself would be a nearly impossible task. But an additional hurdle lies in our way. The fundamental requirements for growing food grains haven't changed much in the past 10,000 years: suitable soil, adequate freshwater, and a climate that food crops can endure. The first two are finite resources, already showing their natural limits as we plow under the last available farmland and steadily drain our aquifers. A benign climate also seems a thing of the past. Though disappearing ice caps and horrific "superstorms" grab the headlines, the largest, most devastating impact of climate change will be on our ability to grow food.

Agricultural experts are already seeing its effects on important crops. The heat wave that hit Europe in 2003 resulted in the deaths of 72,000 people—a headline-making climate tragedy. What most journalists missed, however, was the 20–36 percent plummet in grain and fruit yields that year from crops withered by the heat. In 2010, Russia, one of the world's largest exporters of wheat, lost a third of its crop, thanks to the hottest summer since 1500. Back-to-back droughts in the United States during 2012 and 2013 were the worst since the 1950s, affecting over half the nation's farmland and costing more than $30 billion. The drought-driven food price spikes during 2012 were the third to occur in the previous four years, and they continued to spur social unrest and violence around the world. Just as bread riots ignited the French Revolution, high food prices helped spur the Arab Spring.

The scientific consensus is that we need to limit the rise of global average surface temperatures to 2°C to avoid world-altering climate change. In 2013, our emissions were on track to increase temperatures as much as 3.6°C–5.3°C by century's end. And here's the scariest part: a broad review of climate change studies published by England's Royal Society—the oldest and most prestigious scientific group in the world—concluded that a 4°C increase could make half of the world's current farmland unsuitable for agriculture. The ultimate irony is that agriculture, as currently practiced, is one of the greatest offenders, pumping out a third of the world's annual greenhouse gas emissions. We are literally farming ourselves out of food.

We've actually been in this situation before, but it's not a comparison that gives anyone any comfort. Anthropologists have long pondered why, in the 200,000-year history of *Homo sapiens*, it took so long to develop agriculture, which appeared on the planet only about 10,000–12,000 years ago. Some anthropologists and climatologists now believe that growing crops actually may have been impossible during the previous interglacial periods because of weather extremes. Ice cores in Greenland and lake bed sediments in Europe and North America contain an incredibly detailed climate record of the last 200,000 years, showing large, abrupt spikes of warm and cold periods, wet and dry periods, some of which lasted only a decade or two. Disastrous floods, droughts, and windstorms were much more frequent before the relatively warm, wet, stable Holocene climate of the last 11,000 years that enabled farming to flourish. The not-so-subtle warning is that increasing climate volatility caused by human-induced climate change could, at some point, end agriculture as we know it. And, as one of my agronomy professors liked to say, there can be no culture without agriculture.

AS I WALKED across the eroded fields of Malawi, saw the effects of Punjab's pesticide-poisoned groundwater, and toured massive hog farms in China, the agronomist stirred within me. It became abundantly clear that the 7.3 billion of us who share this incredible planet are intimately connected by our dependence on the soil, the waters, and the climate that feeds us. And unless we put forth a global effort to change our trajectory, these primal elements that have enabled our species to flourish and dominate the planet will not sustain us much longer.

Producing food for more than 9.7 billion people without destroying the soil, water, oceans, and climate will be by far the greatest challenge humanity has ever faced. It will affect everyone, from poor farmers in Africa to the well-heeled suburban grocery shoppers of the West. The fate of the world's great ecosystems, from the

Amazon rain forests to Africa's Serengeti Plain, equally hangs in the balance. The debate over the future of farming is already raging. Advocates of chemical agriculture say we need more technology, better genetics, and better chemicals to meet rising demand. Advocates of organic agriculture argue instead for more diversity, as well as methods that sequester carbon, reduce food waste, and help the world's small farmers be more productive.

Either way, agriculture must change. None of our current agronomic systems have shown much capacity for weathering the vagaries of even the half degree of temperature rise that we've experienced thus far, much less the climate that is forecast to be hovering over our fields in a few short decades. An increase of at least 2°C before the end of the century now seems inevitable. If we hit 4°C, humanity's future looks bleak. Large swaths of the globe may no longer be able to sustain their populations, forcing tens—if not hundreds of millions—to seek refuge elsewhere, spurring more violence and political conflict.

It's not hard to imagine an aging, wealthy, and heavily armed Fortress Europe or Fortress America inhabited by 20 percent of the world's temperate agricultural "haves" attempting to wall off the remaining 80 percent of the world's population—the young, poor, tropical and subtropical agricultural "have-nots," who will do their best to get in. Barring a dramatic decline in population growth, a rapid decrease in greenhouse gas emissions, or a global outbreak of vegetarianism—all of which are trending in the opposite direction at the moment—we're facing nothing less than the end of plenty for the majority of Earth's people.

This is the story of the race to feed the world without wrecking it. It will show you how we got ourselves into this predicament and introduce you to a host of innovative and courageous farmers, researchers, and entrepreneurs who are devoting their lives to getting us out. It will take you from the high-tech genetic labs of

UC Davis, where researchers are developing rice that can withstand floods, to the dusty fields of Punjab, where farmers are using
peasant-based technology to double rice yields with half the water.
It will take you from the giant hog farms of China to the largest
organic farm in the world. It will introduce you to a young Stanford engineer selling low-tech drip irrigation to poor farmers, and a
young aquaculturist launching the largest offshore fish farm on the
planet. It will tell the story of our agricultural past so that we may
learn from our mistakes and—with all of us working together—
create a better agricultural future.

First, however, we must delve into the twin yin-and-yang forces
that are forever at play in the struggle to feed the world: our formidable ability to wrest food from nature, and the unstoppable
"passion between the sexes."

Portrait, by John Linnell, of Rev. T. R. Malthus in 1833, when Malthus was professor of history and political economy at East India College.

CHAPTER 1

The Curse

The superior power of population is repressed, and the actual population kept equal to the means of subsistence, by misery and vice.—*Thomas Robert Malthus, 1798*

The quest to feed a growing population is as fundamental to the human condition as breathing or walking upright. *Homo sapiens* migrated out of Africa after a population explosion that began 70,000 years ago, likely hunting for new sources of food. After the first farmers domesticated plants in Mesopotamia, humans as a species took off. Even rudimentary agriculture could feed 50 times more people than most types of foraging, allowing for larger communities, the leisure to make art or war, and increasing innovation and trade.

Population growth and agricultural production have been locked in a never-ending tango ever since. For the first 4 million years of human existence, our population grew and fell in relation to our food supplies with roughly the same S-curve shown by almost every other species, from bacteria to blue whales. Only after we began farming did our curves start to radically diverge from those of our fellow species. Every agricultural advance—the domestication of plants, the domestication of livestock, irrigation, wet-rice cultivation, the use of legumes in crop rotations—has led to an increase in population, if not an outright explosion. When populations outgrew their ability to feed themselves, hunger exacted its toll until a new food source was found, technology improved, new lands were conquered, or the number of mouths to feed declined.

Numerous early scholars noted the relationship between population growth and available food supplies, including Confucius, Plato, and the author of Ecclesiastes. The latter wrote, some 2,500 years ago, "When goods increase, they are increased who eat them." Even the great classical economist Adam Smith observed in *Wealth of Nations* (1776) that "every species of animals naturally multiplies in proportion to the means of their subsistence, and no species can ever multiply beyond it."

The relationship between our numbers and our food resources is so ingrained in our evolution that it's probably etched in our genetic code. Birth rates typically plunge during famines, whether they occur in Bangladesh, Ethiopia, or Ireland. An estimated 315,000 births were averted during the Irish Potato Famine. But it wasn't until the very end of the eighteenth century that a young British scholar attempted to describe exactly how that relationship worked. For his troubles, Reverend Thomas Robert Malthus became the father of modern demography—the study of populations—and one of the most hated scientists in history.

On an autumn day raw enough to color the cheeks of the ruddiest Englishman, I hopped on the London tube to visit the British Library at St. Pancras and see for myself the book that scholars have been arguing over for nearly two centuries. The librarian in the rare-book collection handed over a thin volume covered with small blobs rimmed by red and blue membranes, looking for all the world like cells dividing. The blurred letters on the cracked red leather spine read simply, "Essay on Population," a truncation of the unwieldy title *An Essay on the Principle of Population as It Affects the Future Improvement of Society, with Remarks on the Speculations of Mr. Godwin, M. Condorcet, and Other Writers*. The author was anonymous, and it was dated June 7, 1798. The meat of the argument appears in the first chapter:

> I think I may fairly make two postulates. First, that food is necessary to the existence of man. Secondly, that the passion between the sexes is necessary, and will remain nearly in its present state . . .
>
> Population, when unchecked, increases in a geometrical ratio. Subsistence increases in an arithmetical ratio. A slight acquaintance with numbers will show the immensity of the first power in comparison of the second. By that law of our nature which makes food necessary to the life of man, the effects of these two unequal powers must be kept equal. This implies a strong and constantly operating check on population from the difficulty of subsistence. This difficulty must fall somewhere; and must necessarily be severely felt by a large portion of mankind.

We now know the writer was a mild, unmarried, 32-year-old curate of a small rural chapel in Surrey who was still living at home with his parents and unmarried sisters. But he was far from a simple country vicar. Robert Malthus (he was never "Thomas") was an interesting man who lived during tumultuous times. He was born in 1766 in nearby Dorking, to a family of modest wealth and radical ideas. His father, Daniel Malthus, was a friend and admirer of philosophers David Hume and Jean-Jacques Rousseau. The second son of seven siblings, young Robert excelled in his studies and was accepted into Jesus College at Cambridge to prepare for a career in the Church of England.

Despite a cleft palate and a harelip, Malthus cut a dashing figure on campus. He wore his blond curls long instead of in a wig as was the fashion, and he spent his free time riding horses, hunting, and voraciously reading the leading thinkers of his day, from Newton to Adam Smith. His friends and the students he later taught remembered him for his kindness, sweet disposition, and good humor. Malthus showed obvious academic talent, especially in mathematics. But the master of Jesus College tried to dissuade him from a career in the clergy because he felt Malthus's speech problems would prevent him from rising through the ranks of the Church of England. Malthus assured the clergyman that all he needed was a simple country church to be happy. Fate, however, intervened. He graduated in 1788 as Ninth Wrangler in Mathematical Tripos—in the top ten of his class—and earned a fellowship at Cambridge just as a series of bad grain harvests and rising food prices were ravaging the impoverished citizens of France. The resulting bread riots, led by women who demanded "just" prices from the bakers, helped ignite the French Revolution, which set ablaze the philosophical landscape of Europe.

Two of the most influential writers of the 1790s were the Marquis de Condorcet, the intellectual father of the French Revolution, and English anarchist William Godwin. They believed that the overthrow of the French monarchy and the spread of representative democracy to the United States were the logical next steps in the inevitable improvement of human society, the natural progression of the Age of Reason that would ultimately lead to a utopian world. Condorcet, in his best-selling *Outline of the Intellectual Progress of Mankind* (1795), argued that advances in reason, science, and technology would ultimately create a perfect, harmonious human society, without the need of governments or monarchies. Godwin, publisher of the periodical the *Enquirer*, believed all institutions morally corrupt, including orchestras and marriage. Rid the world of these unjust organizations, and a benevolent world of common property (including wives) would follow, where fields produced an increasing bounty, the desires of the flesh would be replaced by the exercises of the mind, and humans would live longer and longer until they achieved God-like immortality.

Malthus's father had read one of Godwin's essays in the *Enquirer* and was inclined to agree with him. Young Robert took a far more pragmatic view. England was in the midst of a population boom at the time, though few were aware of it until the first census in 1801. As an acute observer of rural poverty around his small country church, Malthus couldn't help but notice that the number of christenings far exceeded the number of funerals. Large landowners were enclosing the common lands, denying cottagers fields for their livestock and the vegetable gardens that helped them keep food on the table. Instead of the plump, cherry-cheeked innocents beloved by landscape painters of the day, he saw scores of hungry peasant children, stunted from malnutrition. And despite Godwin's loathing of the monarchy and the church, the bloody Reign of Terror that followed the French Revolution and took Condorcet's life made England's "corrupt" institutions seem pretty sound by comparison.

Such observations inspired Malthus to dash off a response to his father, as well as to Godwin and Condorcet, that respectfully guillotined their utopian ideals. As a mathematician and lover of statistics (he would later cofound the group that became the Royal

Statistical Society), Malthus turned to the most solid data set he could find: Ben Franklin's remarkably accurate estimates of population growth in the young United States, where fertile land and food resources were abundant.

According to Franklin, the population of the former British colony had been doubling about every 25 years—lower than its maximum potential, Malthus reasoned, but good enough for his argument. In America, the population grew quickly because it had abundant game, fisheries, and most important, unlimited lands that could be brought into production. But in old established countries of Europe, extracting more food from the land was much more difficult. By breaking up forests and pastures into tillable fields and applying more manure as fertilizer, Malthus calculated that England's agricultural production might double in 25 years. Even if England could continue increasing its production by the current amount every 25 years, however, the increase would still be arithmetical. In other words, Malthus argued, if y was the yield of all of Great Britain, it would increase from $1y$ to $2y$ to $3y$ to $4y$ every 25 years, while the population (call it p), which Franklin had shown could double every 25 years, would increase from $1p$ to $2p$ to $4p$ to $8p$ and so on. Within a century, more than half the English would have no food.

Modern critics have argued that Malthus predicted the world's population would ultimately balloon and then crash—the famed "Malthusian collapse." But Malthus was simply demonstrating the wide chasm between logarithmic and arithmetic growth. Instead he tried to explain why populations didn't collapse, but rather fluctuated in step with their food supplies. Enough tree seeds fell in England every year to forest five new worlds, he said. But only a fraction survived, because there was not enough soil, water, or space for all.

Malthus believed that human populations were no different.

There had always been, and always would be, powerful checks on human numbers. The overarching check was the food supply, but only during severe famines was hunger the direct cause of death. Such disasters occurred often enough in his day. The Bengal famine of 1770 killed an estimated 10 million people—a third of the population of Bengal—under the callous watch of the British East India Company.

The more common everyday checks he placed into two broad categories. "Positive" checks were anything that shortened human life and increased the death rate. These included severe or unhealthy labor, exposure to the elements, extreme poverty, poor nursing of children, and "the whole train of common diseases and epidemics, wars, pestilence, plague and ultimately famine." He even included living in great cities such as London, which, with its open sewers and cholera-plagued slums, had far higher death rates than the rural countryside. The Great Plague of London in the 1660s killed an estimated 100,000 people, roughly 15 percent of the city's inhabitants. In Malthus's day, life expectancy was only 40 years.

Anything that reduced the birth rate Malthus called "preventative" checks to population. Men could see the struggles of large families, look at their week's pay, and make rational calculations about how far the money might go among seven or eight kids. When times were hard, single men in England and Europe typically held off marrying and starting families until they felt they could support them. Delaying marriage invariably reduced the number of children a couple had, eventually reducing the labor supply, which caused wages to rise, enabling men to marry earlier, in endless oscillating cycles. Almost anything that affected the price of food or real wages affected the ability of a poor laborer to marry and strengthened or weakened the preventative check, including wars, disease outbreaks, economic booms or depressions, and agricultural advances. Even government policies such as

duties on imported food or Poor Laws that helped feed the indi-
gent could affect population growth. Should all other checks fail,
Malthus wrote ominously, "gigantic inevitable famine stalks in the
rear, and with one mighty blow levels the population with the food
of the world."

Malthus believed such cycles had been hidden through history
because most historians chronicled the upper classes, while the
great burden of population checks fell on the poor. But preventative checks were at work at all levels. Gentlemen of modest means
and learning, such as Malthus himself, often chose to have fewer
children or not to marry at all, since the cost of a family might
lower their status in society and take money away from more lei-
surely pursuits. Malthus didn't marry until he was 38, after he'd
secured a stable income from the church, and he had only three
children—half the average at the time. Even servants who made a
comfortable living for one, he reasoned, might not want to divide
their wages among four or five. Either way, the twin pressures of
population and food supply made Condorcet and Godwin's utopia
sheer fantasy:

> The natural inequality of the two powers of population and
> production in the earth . . . form the great difficulty that to
> me appears insurmountable in the way to the perfectibility of
> society. . . . I see no way by which man can escape from the
> weight of this law which pervades all animated nature. No fan-
> cied equality, no agrarian regulations in their utmost extent,
> could remove the pressure of it even for a single century.

The idea that the primary cause of misery and poverty among
the working classes was an immutable natural law brought on by
their own fecund loins provoked a firestorm of protest. But it was
Malthus's moral applications of the law that probably riled his ene-

mies the most. The misery caused by the population principle, he reasoned, was part of God's grand design. The fear of hunger or falling in social stature compelled men to work hard, support their children, avoid their vices, and lead more virtuous lives.

Though he always believed in short-term relief for the aged, infirm, and distressed, Malthus argued that England's Poor Laws, which entitled a poor man to a certain amount of money for each child, should be gradually phased out. By removing the threat of hunger, the government was encouraging able-bodied men to marry early and have children they could not feed—a process that led to misery and sin in Malthus's view. The Poor Laws, he said, simply created more poor people, raised food prices, and depressed real wages without increasing food supplies by one bushel. An able-bodied man who cannot find work, Malthus later wrote, "has no claim of *right* to the smallest portion of food, and in fact, has no business to be where he is. At nature's mighty feast there is no vacant cover for him." It was tough love, eighteenth-century style. And Malthus was excoriated for it.

Karl Marx called him "the great destroyer of all hankerings after human development," painting Malthus as a defender of the landed gentry and the corrupt status quo. In socialist societies, Marx argued, Malthus's principle of diminishing returns in agriculture was utter hogwash: every comrade could raise enough food for himself, ad infinitum, making socialist countries capable of supporting any population. Marxists and socialists continue to be Malthus's most virulent critics.

William Godwin railed back at Malthus like a stung cat, arguing that humans are indeed masters of their own fate and with our "ingenuity" can overcome any problem, including overpopulation. Even conservative Tories like William Cobbett declared early marriage to be the "the greatest of all compensations for the . . . hardships and sorrows of life." But it was the humanitarians and social reformers of his day that painted Malthus as the callous enemy of the poor. Although Malthus was a lifelong advocate of universal public education, the extension of civil liberties to the landless classes, land reforms, and shifting government emphasis from luxury-goods manufacturing and foreign trade to agriculture in order to produce more food for people, he was seen as the man whose "natural law" undermined any attempt at government reform. Even Charles Dickens, whose father was sent to debtor's prison, savaged Malthus in *A Christmas Carol* by making Scrooge a heartless Malthusian. When men come to his door to collect alms for the poor, the miser tells them that the unfortunate pauper who would rather die than go to the workhouses "had better do it, and decrease the surplus population."

Despite the pillorying he took from critics, Malthus remained hugely influential. He became great friends with rival economist David Ricardo, who disagreed with Malthus on many points but never questioned his population theory. Malthus's letters to Ricardo debating the theory of rent, tariffs on imported corn, and their concurrent discovery of the law of diminishing returns in agriculture are still read by economics students. Malthus became the first professor of political economy in England at Haileybury College, established to train the future officers of the East India Company. While there, he published the *Principles of Political Economy Considered with a View to Their Practical Application* in 1820, one of the fundamental texts of classical economics. He had enormous influence on John Stuart Mill and later John Maynard Keynes, who put him on the same pedestal as Locke, Hume, Adam Smith, and Mill. Even Charles Darwin and Alfred Russel Wallace credited Malthus's population principle, with its constant pressure from subsistence, with giving them the key to understanding the evolution of species.

But it was that same principle—particularly the slow arithmetic growth of food production—that made Malthus the whipping boy of late-twentieth-century economists. Around the time Malthus was writing in the early 1800s, England was undergoing a dramatic agricultural revolution. The peasant system, little changed since Roman times, soon metamorphosed into the precursor of modern, industrial agriculture. Breeders were busy developing new types of livestock, such as Robert Bakewell's fat, fast-growing New Leicester sheep, which he reportedly dubbed "machines, for turning herbage . . . into money." Other gentlemen farmers were experimenting with crop rotations like the Norfolk method, brought over from Flanders, which incorporated turnips and clover into rotations of wheat, flax, barley, and corn. The turnips and clover were fed to cattle in pens so that the manure could be collected more efficiently for use as fertilizer. The new rotations meant fewer years when fields were left fallow, increasing food production.

Most important, farmers discovered that legumes like clover and alfalfa miraculously restored worn-out soil. In Malthus's day, as well as today, nitrogen was the critical limiting nutrient in grain crops. Legumes, such as beans, peas, and clovers, have tiny rhizobium bacteria living in little nodules on their roots that can convert atmospheric nitrogen (N_2) into ammonium (NH_4) that the plants can use—possibly the most useful symbiotic relationship on Earth. Grains like corn, wheat, and rice are members of the grass family. They don't have nodules or deep root systems, but instead throw all their energy into making seeds—one reason why they produce far more edible seeds per acre than legumes do, as long as they have sufficient nitrogen.

Agricultural historians in England estimate that in the county of Norfolk alone, the introduction of clover and other legumes in crop rotations tripled the rate of nitrogen fixation in the area. As a result, wheat yields grew by a quarter between 1700 and 1800, and again by half between 1800 and 1850. Farmers also began using marl—the precursor to agricultural lime—to reduce soil acidity. And 200 years before the rock band Jethro Tull began tearing up England with flute-driven rock 'n' roll, their namesake, the agricultural inventor Jethro Tull, was tearing up the English countryside with his new horse-drawn hoes, plows, and seed drills, launching a mechanical revolution in agriculture that reduced labor, increased efficiency, and enabled farmers to tend far more land. Tull's seed drills planted crops into uniform rows 36 inches apart—the same basic technique used on my family's North Carolina farm for corn, tobacco, cotton, and soybeans until the mid-1980s.

The resulting boost in the yield per acre coincided with a rapid increase in the number of acres under production. Forests, hilly heaths, and lowland fens that had once been considered wastelands were ditched, drained, and planted. At the same time, rising crop prices and the greed of England's landed aristocracy led to the final enclosure of the British commons, a Parliament-sanctioned landgrab that had been under way since the 1400s. The Enclosure Acts took away the common lands that rural peasants used for subsistence agriculture and put vast amounts of land in the hands of a few. Today, 0.6 percent of the British population owns half of the nation's rural land—among the highest concentrations in the world. A full third is owned by just 1,200 aristocrats.

THE BRITISH AGRICULTURAL revolution that occurred between 1750 and 1850 produced so much food that the population of England nearly tripled, from 5.7 to 16.6 million. More important, the enclosure of the commons and the loss of farmwork to machines drove so many laborers off farms and into towns and factories seeking work that it laid the foundation for the industrial revolution that followed. Malthus was ten years old when the steam engine was invented. He had no idea how that simple, hissing, smoke-belching machine would utterly transform the world, or what would happen when its coal-powered might was harnessed to the plow, or the threshing machine, or the factory floor.

The biggest charge against Malthus was that his calculations didn't consider the potential for transformational technological change. But none of the classical economists, Malthus included, foresaw the great industrial revolution or the phenomenon of modern economic growth based on coal and other fossil fuels—the stored energy of 300 million years of photosynthesis. Malthus could not imagine the improvements in medicine and sanitation that would cause the death rate to fall, the widespread use of contraceptives to prevent births, or the chemical fertilizers and hybrid seeds that would enable farmers to feed the spiraling population.

In defiance of Malthus's principle, world population nearly quadrupled from 1.6 to 6.1 billion from 1900 to 2000—the largest increase in human history (though still far below Malthus's benchmark for unchecked populations doubling every 25 years). In that time, world grain production increased fivefold, from 400 million to 1.9 billion tons—or as agricultural economist Lester Brown puts it, five times more in one century than we were able to produce in the previous 10,000 years combined.

Malthus's legitimacy was further tarnished in the early twentieth century, when advocates of eugenics and social Darwinism used his words to justify their abhorrent movements. His population theory briefly came back into vogue in the 1960s, when spiraling population growth and world grain shortages led several writers—including Stanford biologist Paul Ehrlich (*The Population Bomb*, 1968), ecologist Garrett Hardin ("The Tragedy of the Commons," 1968), and the European think tank the Club of Rome (*The Limits to Growth*, 1972)—to predict global famine and population collapse. These "Neo-Malthusians," as they were called, lost credibility during the green revolution, when the rapid increase in agricultural production not only defused the population bomb but led to a global glut of grain during the 1980s and 1990s. "Malthus has been proved wrong time and time again," Cornell agricultural economist and development expert Christopher Barrett told me, echoing his peers.

Yet during the consecutive food crises between 2008 and 2012, Malthus was everywhere—the *New York Times*, the *Economist*, *Foreign Policy*, among many others. Why, if Malthus was so wrong, is his theory so hard to dismiss?

"Most of the people who say Malthus is wrong haven't read him. And they aren't being fair to him if they have," says Timothy Dyson at the London School of Economics, one of the world's most respected demographers. "For example, he didn't predict that there would be famines in the future or that the world would run out of food. He wasn't looking at the global picture to begin with. . . . Part of the reason why we go back to him is simply that no matter which culture you're in in the world, whether European, Chinese, or Arab, there has been someone who has basically said the same thing."

Most Malthus haters attack his youthful first essay of 1798. Few of his critics mention the complete rewrite he did in 1803 after extensive travels in Scandinavia. On his five-month-long trip, he gathered data on local populations, the size of families, the average age at marriage, the price of rents, labor, and food; and he spoke with prominent scholars along the way. The second edition was essentially a new book, expanded from 50,000 to 250,000 words, and adding numerous examples of population growth in Europe and other parts of the world based on the best data he could find. He revised the book with new numbers four more times before his death, and by 1806 he had deleted the most offensive passages and softened his harsher conclusions about the future of humanity. Though his population principle was left unchanged, Malthus observed how quickly human numbers recovered after wars or famines and realized that falling birth rates had a far greater impact on populations than did rising death rates.

Thus, humans could control their own numbers through a third check, which he called "moral restraint." By this, Malthus meant simply voluntarily delaying marriage and observing abstinence out of wedlock.

In the mid-1980s, two of England's top historical demographers—Sir Anthony (Tony) Wrigley and R. S. Schofield at Cambridge published the painstakingly researched tome *The Population History of England, 1541 to 1871*, replete with data on age of marriage, birth rate, death rate, rate of population growth, and real wages, among other indicators. When they applied Malthus's second (1803) model based on delayed marriages and real wages to the historical record, they discovered that he had accurately described the relationship between population and resources in England, as well as much of western Europe, for the 300 years prior to the industrial revolution. Few social scientists in history have ever proposed a theory that has held for three centuries.

Perhaps the real reason we can't shake Malthus, however, has more to do with the human psyche than with demographic theories. In reconstructing the populations of the past, historical demographers have found numerous instances of populations that *have* grown past their resource base and then been cut back down by horrific famines. Over the centuries this phenomenon occurred in France, Japan, Egypt, India, and China. Perhaps that is why the Chinese word for "population" consists of two characters—one representing a person next to another showing an open mouth. Malthus reminds us of our primal fears—starvation, hunger, and want.

"Ultimately there has to be a balance between population and resources," says Dyson. "And this notion that we can continue to grow forever—well, it's ridiculous. There's been an increase of more than 4.3 billion people in my lifetime. And just because the world's population has somehow been able to live within its means and improve our levels of living doesn't mean that it's not going to have huge consequences."

Hungry peasants await the opening of a government grain shop in Calcutta during the Bengal famine in November 1943.

CHAPTER 2

Famine's Lethal Lessons

A series of calamities, each one of unprecedented magnitude, followed in such quick succession that the administration was overwhelmed. It was a "Dunkirk" on the food front in Bengal.—*M. Afzal Husain, 1945*

On the morning of October 16, 1942, Samar Sen, a submagistrate in the Indian city of Contai, was at work as usual in his third-story office. Luckily for the administrator, he was in one of the newer cement buildings in the ancient agricultural town on the Bay of Bengal. The weather had been rainy and windy all morning. By 11:00 a.m. the gusts were so strong that Sen could not keep his door shut. Then the windows blew out. For the next 12 hours the building became his storm shelter, as one of the strongest cyclones ever recorded—with estimated 200-mile-per-hour winds and 20-foot seas—laid waste to his city and the surrounding rice fields. He later described the devastation he saw the following morning:

> Not a hut was standing within the miles of Contai. Fields were completely under water . . . and dead bodies were floating on the fringe of the town. . . . Death toll had apparently been heavy but nothing could be seen on the countryside which had gone completely underwater. . . . Crops, so beautiful and good only 24-hours ago, had completely vanished and it was obvious much food grains had been washed away.

The damage wasn't limited to Contai. Three successive tidal waves swept across the region, flooding 3,200 square miles of the surrounding district of Midnapur—an area larger than Delaware and Rhode Island combined. A subsequent government report listed 14,500 killed, as well as 190,000 cattle, with dwellings, food stores, and crops destroyed over a wide area.

Midnapur was the major rice-growing district in undivided Bengal, a densely populated region home to 60 million people that at the time included modern Bangladesh. The cyclone struck the *aman* crop just a few weeks from harvest. The largest and most important of Bengal's three yearly rice crops, the *aman* provided almost three-quarters of the province's annual grain production. Most Bengalis at the time were either poor farm laborers with no land of their own, or small-scale subsistence farmers who were only slightly better off. Rice—nutritious, delicious, and cheap—was the bedrock of their diet. It had fed peasants and Mogul princes for more than 3,000 years.

"Rice is vitality, rice is vigor too, rice is the means of fulfillment of all the ends of life," reads a passage from the *Krishi-Parashara*, an ancient Sanskrit text. "Gods, demons and human beings all subsist on rice."

This was not necessarily a good thing. By 1942 Bengal was more dependent on rice than the Irish were on potatoes prior to the 1845 blight. Even though more than 90 percent of Bengal's fertile land was planted in the crop, the state still needed to import more than a million tons of grain each year from Burma and other parts of India to make sure there was enough for all. Bengal's average harvest of 9 million tons amounted to only a pound a day for each adult male and even less for women and children—just enough for a meager existence.

World War II made the food situation dire. The Japanese army routed British and Chinese forces in neighboring Burma in the spring, cutting off all rice imports to India and sending a flood of soldiers and refugees into eastern Bengal. Fearing a Japanese invasion, the British military requested Bengal's nominally independent government to impound all rice stocks to feed vital war workers in Calcutta, and to destroy more than 60,000 small boats in the coastal regions to keep them out of Japanese hands. This "boat denial policy" made moving rice from coastal to inland markets difficult, not to mention putting thousands of poor fishermen and boatmen out of work. To make matters worse, some of the rice crop that year had been infected with brown spot, a virulent fungal disease that could cut yields in half. The cyclone's powerful winds spread the brown spot spores far beyond the flood zone. Two government rice research stations (one 200 kilometers northeast of Contai and the other 176 kilometers northwest of the city) reported yield losses of 40–90 percent. By some estimates, a third of Bengal's rice crop was lost.

The market responded with a vengeance. The price of rice, which had been slowly rising since the summer of 1942, suddenly doubled, fueling hunger marches and protests. In February 1943 the agricultural minister of Bengal declared a major shortfall, spurring panic buying in the markets. By midsummer, rice was selling for six times its 1942 price. Rural laborers who could barely feed their families in good years began reducing their meals. As their bellies swelled and their skin turned orange, they started eating grass and leaves. They stole grain from fields and stores, and protested in the streets of their villages. Small farmers began selling their land just to buy food. Then they started selling their wives and children.

By March of 1943, ruthless famine stalked the land. It started in the cyclone-struck areas of West Bengal and the rural areas of East Bengal—now Bangladesh—that typically required food imports. By midsummer it had spread throughout the region, and an army of emaciated zombies began shuffling into Calcutta to beg for handouts. They stood in long lines in front of government-subsidized food shops, defecated on the streets, caught dysentery and other diseases. Then they began to die.

Newspapers of the day printed ghastly images: mothers clutching dead babies to their shriveled breasts; babies trying to nurse their dying mothers; an emaciated boy lying in the dust beside his emaciated dog; a dog gnawing on a human skeleton. People died by the hundreds, then by the thousands as disease and dehydration set in. By the time a huge *aman* crop finally broke the famine in the late fall of 1943, more than 2 million Bengalis had perished. Some experts put the death toll far higher. Many died next to lush green rice fields just weeks away from the harvest that might have saved them.

THE BENGAL FAMINE of 1943 was one of the most catastrophic, controversial, and studied food disasters in modern history—one that nearly 70 years later still ignites heated debate from the streets of Calcutta to the ivory towers of Harvard. It tarnished the legacy of Winston Churchill with cries of genocide and racism for failing to rush relief to the stricken corner of the British Empire, and it became a rallying cry for the Indian independence movement. The disaster's legacy spread far beyond the Indian subcontinent, however, ultimately inspiring two Nobel Prize winners who had a profound impact on the global food debate. One radically changed our understanding of why famines occur. The other transformed global agriculture in his quest to prevent them.

By the 1940s, industrial agriculture had undermined Malthus's theories on food production. But his view on the cause of famine had endured. In his second, revised essay of 1803 he wrote:

> But though the principle of population cannot absolutely pro-duce a famine, it prepares the way for one in the most complete manner; and, by obliging all the lower classes of people to subsist nearly on the smallest quantity of food that will support life, turns even a slight deficiency from the failure of the seasons into a severe dearth; and may be fairly said, therefore, to be one of the principal causes of famine.

Bengal, in particular, was no stranger to such catastrophe. When Malthus was a boy, successive droughts led to the epic famine of 1770, which was followed a century later by a series of famines between 1860 and 1877 that led to the development of the British Famine Codes—a handbook for British officials dictating the orderly provision of relief to the starving. During the 1800s, India, China, and Ireland all suffered devastating famines that killed millions to tens of millions of people. Though the actual data on such historic death tolls are notoriously poor, economic historian Cormac Ó Gráda at University College Dublin has conservatively estimated that nineteenth-century famines killed 100 million people—30 million from India and China alone during the last three decades. That's 8 percent of the average global population during the 1800s.

Even the twentieth century, with its increasingly industrial agriculture, saw famine on an epic scale. The Soviet Union lost 15–19 million people during three brutal multiyear famines that began in 1921, 1931, and 1946. Ukrainians call the 1932–33 famine the *Holodomor*, which translates into "extermination by hunger." Many still accuse Stalin of crushing the Ukrainian nationalist movement through famine-induced genocide that killed an estimated 5 million people, though deaths occurred in other areas as well. China suffered similar famines in 1927, 1929, and 1943 that killed between 2 and 6 million people each. But those paled by comparison to the starvation that occurred during Mao's Great Leap Forward. The famine death toll between 1959 and 1961 has been estimated at an unimaginable 30 million Chinese, with another 33 million in lost or postponed births, as drought and bleak harvests coincided with the Chairman's disastrous attempt to revolutionize Chinese agriculture.

As Malthus noted, famines rarely hit all segments of society, but instead fall heaviest on those who have the least to lose. This phenomenon fascinated economist Amartya Sen, who was a young boy living in the small university town of Santiniketan in Bengal during the 1943 famine. As the son of academics who taught at the university, he barely noticed it. "I knew of no one in my school or among my friends and relations whose family had experienced the slightest problem during the entire famine," Sen once wrote. "It was not a famine that afflicted even the lower middle classes— only people much further down the economic ladder, such as landless rural labourers."

That revelation stayed with Sen as he studied economics and philosophy at universities in Calcutta and England, eventually earning his PhD from Cambridge. Though his first love was social choice theory—the study of collective decision making ironically pioneered by Malthus's philosophical nemesis, the Marquis de Condorcet—Sen became interested in how such theories could be used to address real-world problems like poverty, inequality, and gender disparities. During the mid-1970s, while teaching at the London School of Economics and at Oxford, Sen began investigating the cause and prevention of famines. While the former British administrators of India, trained in the Malthusian tradition, focused on the macro issue of gross food supply available per person in a region, Sen saw famines as economic instead of agronomic disasters. He looked instead at the microeconomics of the poor and their ability to buy or otherwise acquire food during such crises.

Though he felt that poor harvests could contribute to famine, Sen became convinced that many famines occurred when there was plenty of food to go around. In these cases people starved when they lost what Sen calls their "food exchange entitlement"— their ability to acquire enough food when they lost income, when they no longer had assets to sell or trade, or when food prices rose faster than their wages. Such inflation was often spurred by inept government policies, speculation, or hoarding, rather than any significant decline in food. Even perceived shortages, Sen believed, could lead to panic buying, driving prices out of reach of the poor.

In *Poverty and Famines: An Essay on Entitlement and Deprivation*, published in 1981, Sen reexamined crop production data in four major twentieth-century famines: Bengal, 1943; Ethiopia, 1972–74; Bangladesh, 1974; and the Sahel, 1968–73. According to his calculations, only in the Sahel famine did food availability decline significantly, yet even in that case, Sen argued, knowing the overall food supply revealed little about who had starved or why. Natural disasters like cyclones, floods, or droughts were often simply convenient scapegoats that corrupt or feckless politicians blamed for the deaths and devastation that occurred on their watch.

The great Bengal famine of 1943, Sen said, was a classic example. After the disaster, British Viceroy Lord Wavell appointed a Famine Inquiry Commission to investigate the cause. The blue-ribbon panel included some of India's top experts on nutrition and agriculture, and in 1945 they placed the blame squarely on the cyclone, brown spot disease, and the loss of imported rice from Burma that caused "the serious shortage in the total supply of rice available for consumption in Bengal." But using the panel's own crop production figures, Sen calculated that the supply of rice available in 1943 was actually 13 percent more than in the previous poor crop year of 1941—a year in which there had been no famine.

Instead, Sen blamed the millions of deaths on hyperinflation caused by the war boom and poor government policies, especially the overprinting of currency to finance the war effort. Inflation fueled rampant speculation and hoarding by traders that quickly drove the price of rice out of reach of the poor. Sen called this a "boom famine." The Bengal government's boat denial policy, along with its decision to purchase rice at any price on the open market to provide food for the Indian army and a million war industry workers in Calcutta, exacerbated the problem. Millions died not from crop failure, but from the failure of the government and the market to distribute the crop equitably. Thus the great Bengal famine of 1943, Sen said, was entirely "man-made."

Sen's theory was so influential and widely praised that it replaced Malthus's food shortage theory overnight. After Sen, nearly every famine, both modern and historic, has been viewed through the "man-made" lens, no matter how bad the drought or flood or harvest. In later publications Sen went even further, asserting that no major famine had occurred in a functioning democracy with a free press, where elected officials could be held accountable for their inaction. "Famines are, in fact, so easy to prevent," Sen once wrote, "that it is amazing that they are allowed to occur at all." Instead of giving the starving food, he suggested, in many cases they would be better served if governments gave them money to buy food on the open market.

For his vast body of work on social change theory and welfare economics, which focused the world's attention on the plight of the poor, Sen was awarded the Nobel Prize in 1998.

There is a lingering problem with Sen's theory, however. From the earliest papers he published on the Bengal famine in the 1970s, numerous scholars have challenged the cornerstone of his argument—namely, that there was more food available in Bengal in 1943 than in the nonfamine year of 1941.

Sen based his conclusion primarily on data published by the Famine Inquiry Commission (FIC) in 1945, which analyzed crop production estimates, population estimates, and per capita nutrition requirements for the 15 years preceding the famine to estimate the food supply in 1943. But a key piece of the data was missing: the amount of rice stocks carried over from previous years. Political leaders and agricultural economists carefully track these numbers today because of their powerful influence on food prices. But in the 1940s, only rice-dependent Japan routinely counted such stocks. The FIC estimated that 13–33 percent of Bengal's annual rice production was carried over every year. No famine had occurred in 1941, they reasoned, because the bad harvest had been preceded by several good years that produced large surpluses. But by their calculations, the poor crop of 1941, combined with the loss of imported rice from Burma in March of 1942, damage from the cyclone, as well as an "indifferent" rice crop in the rest of Bengal that year, had left only a six-week supply to start the year. The FIC blamed the sharp fall in carryover stocks as the primary cause of the crisis but admitted that its conclusion was pure conjecture, since no hard data existed and the margins of error were large.

Sen scoffed at this painstaking hunt for a shortage, famously calling it "a search in a dark room for a black cat which wasn't there." A brilliant mathematician, Sen took the FIC's own data from 1938, made his own corrections, and calculated two- and three-year moving averages of food supplies in Bengal between 1938 and 1943. Sen's annual figures showed that Bengal's total per capita food grain supplies were 9 percent greater in 1943 than in 1941. The two-year moving average put 1943 supplies 7 percent higher than those of 1941, while the three-year moving average put 1943 only 1 percent lower than the nonfamine year. "It seems safe to conclude," Sen wrote, "that the disastrous Bengal famine was not the reflection of a remarkable over-all shortage of foodgrains in Bengal."

More important, Sen believed that the Malthusian method of calculating overall food supplies versus the population to gauge food availability had blinded officials in Delhi in early 1943 to the escalating famine building in Bengal from spiraling prices and the loss of purchasing power among the rural poor. Lord Linlithgow, the British viceroy at the time, initially asked Churchill's war cabinet for a mere 600,000 tons of extra wheat from Australia to combat the famine—a minuscule amount compared to India's 60- to 70-million-ton annual production. The request was promptly denied. As the crisis deepened, Linlithgow began begging Whitehall for wheat to feed Bengal. But Churchill was already trying to relieve a famine in Greece and feed armies on three fronts—not to mention losing a merchant ship a day in the region to Japanese destroyers. Instead, the British Bulldog growled about Indians "breeding like rabbits" and asked why Gandhi, his arch rival, hadn't died. He spared no wheat for Bengal.

The FIC report, however, was not unanimous. The sole agronomist on the commission, Professor M. Afzal Husain, included a scathing 20-page dissent claiming that food supplies in 1943 Bengal had been far lower than the estimates reached by the rest of the panel. A native of the Punjab, Husain had earned the top honor at Cambridge in zoology in 1916, and he had gone on to become India's leading entomologist. Husain pointed out that the figures for rice production in Bengal weren't harvest data at all, but rather preharvest forecasts often compiled by illiterate village watchmen and then adjusted by government officials up the line. Some evaluations of this method put the margin of error as high as 150 percent. Even the FIC had raised Bengal's production estimates by a ballpark 20 percent because they seemed too low. "With statistics so hopelessly defective," Husain wrote, "either no attempt at all should be made to evaluate the [food supply] position, or the conclusions drawn from the estimates available should be subjected to various tests and their reliability determined." Sen had based his conclusions on the same dodgy numbers.

According to Husain's arithmetic, Bengal had been in a food deficit every year since 1934, save one (1937). The province's increasing dependence on grain imports from Burma and other Indian provinces belied the notion of any annual surplus, while an intensive government search for hoarded rice—another suspected cause of the famine—turned up little. Increased wartime demand from troops, defense workers, and refugees played a role, Husain wrote, but absolute food shortage was the disease. "High prices, like high temperature, were merely the symptom."

About the only data from the Bengal famine that hasn't been seriously challenged came from another agronomist, S. Y. Padmanabhan, who had been appointed to the position of plant pathologist for Bengal in October of 1943, in the midst of the crisis. On the trip to his new post, Padmanabhan had passed "dead bodies and starving and dying persons all along the way from Bahudurabad Ghat on the Brahmaputra to Dacca," a distance of more than 200 kilometers. Padmanabhan would eventually become director of the Central Rice Research Institute in Cuttack and would play a critical role in increasing rice production during the green revolution in India during the 1970s. On the thirtieth anniversary of the Bengal famine, he published a detailed paper on the epidemic of brown spot (caused by a fungus called *Helminthosporium oryzae*) that had infected the rice crop after the cyclone in Bengal that year. It was Padmanabhan who found the field trials carried out at the two rice research stations in central Bengal and discovered that the disease had resulted in 50–90 percent yield declines.

These were not adjusted guesses made by illiterate chokidars, but data from research plots carefully measured, harvested, and recorded in kilograms per hectare by scientists, students, and trained workers. Padmanabhan then gathered and analyzed weather data for sunlight, humidity, temperature, and rainfall from several other weather-reporting stations in the fall of 1942, and he found conditions across a large swath of Bengal to have been ideal for the spread of the disease. In the 1950s he and his students conducted elaborate studies on the subject, collecting and counting spores of *Helminthosporium* on greased glass plates to determine exactly what those optimal conditions were, so that Indian farmers would be better prepared for blight the next time. Unlike Sen, the brilliant theoretical economist, Padmanabhan was a classic, old-school crop scientist who collected his own data and then field-tested his results.

"Though administrative failures were immediately responsible for this human suffering, the principal cause of the short crop production in 1942 was the epidemic of *helminthosporium* disease which attacked the rice crop in that year," Padmanabhan wrote. "Nothing as devastating as the Bengal epiphytotic of 1942 has been recorded in plant pathological literature. The only other instance that bears comparison in loss sustained by a food crop and the human calamity that followed in its wake is the Irish potato famine of 1845."

It's probably not surprising that a philosopher-economist saw a severe political and economic failure, while agronomists saw blight and crop loss on a biblical scale. Several modern scholars have since agreed with the agronomists. It now seems relatively clear that a true Malthusian dearth of food *and* inept government policies conspired to create the Bengal famine of 1943. Sen deserves enormous credit, however, for shifting the world's focus from natural disasters to famine prevention and the economic lives of the poor, who still remain the greatest victims of both crop and market failures. Shortly after *Poverty and Famines* was published, the US Agency for International Development created the Famine Early Warning System (FEWS) in 1985 to alert the world to drought, flood, and rising food prices, enabling governments and relief agencies to marshal food aid to affected regions before starvation occurs.

The Bengal famine proved to be a watershed event both in the response to famines and in overall world food supplies. It was the last major famine in India. Although other famines followed that killed millions in China, Russia, Cambodia, and North Korea, these sprang more from war, collectivization of farms, and other lethal government policies that either crippled agricultural production or turned poor harvests into national disasters. In the late twentieth century, classic famine shifted almost entirely to Africa, where dearth has often been exacerbated by conflict and civil war. Even so, famine death tolls there have been far lower than those of Asia or Europe, typically measured in tens or hundreds of thousands rather than millions. Part of the reason has to do with the increased vigilance of aid agencies like the United Nations' World Food Programme and the increased monitoring of famine-prone countries. The global public is also more informed than ever. Major television news networks and newspapers are quick to send reporters to cover food crises in places like Ethiopia and Darfur, helping, as Sen argued, to shame governments into action.

Yet one can't overlook two overarching factors that drastically changed the world food equation. In 1981, when Sen published his convincing argument in *Poverty and Famines* that access to food is more important than production, the world was in the midst of a global crop boom unlike any other in history, making more food available to more people than ever before. Moreover, in the second half of the twentieth century, India, China, and the Soviet Union drastically increased the amount of food they produced while cutting their fertility rates (the number of children that women bear in their lifetime) in half. The countries of sub-Saharan Africa did neither.

Bengal harvested record, back-to-back *aus* and *aman* rice crops in the fall of 1943 that finally ended the famine, while the new viceroy sent in the army to ensure that the crop was distributed fairly. Yet India continued to suffer from chronic food shortages, with a third of its population malnourished in any given year. When Jawaharlal Nehru became the first prime minister of an independent India in 1947, he made his priorities clear: "First of all, obviously, we must have food and enough food," Nehru said soon after his election. "Everything else can wait, but not agriculture."

On the other side of the planet, tromping around a wheat field in Mexico, a tough, plainspoken plant pathologist from the American heartland heard Nehru's call. And the world would never be the same.

Norman Borlaug checks wheat plants for rust resistance in 1964 at a field station near Ciudad Obregón, in Sonora, Mexico, the birthplace of the green revolution.

CHAPTER 3

The Green Revolution

Food, Sex, and War

If you desire peace, cultivate justice, but at the same time cultivate the fields to produce more bread; otherwise there will be no peace.—*Norman Borlaug, 1970*

Most people on the planet wouldn't recognize the man in the old black-and-white photo. But to agricultural researchers, the picture of Norman Borlaug is as iconic as that of the marines lifting the Stars and Stripes over Iwo Jima. He stands in the middle of a Mexican wheat field in 1964, his khaki work pants obscured by wispy heads of grain that reach nearly to his belt. His shirtsleeves are rolled high, revealing a brawny, sun-darkened right arm that scribbles in a large notebook. His trademark straw trilby is cocked back on his head, his eyes steeled on the wheat field in front of him, judging its height, vigor, and yield.

Within a few decades Borlaug and his fellow researchers would do more to vanquish world hunger than anyone before or since. Even though the number of people on the planet has nearly tripled since 1950, the high-yielding seeds and chemical-intensive farming systems they pioneered have helped slash the number of chronically malnourished from about 30 percent of the global population in 1950 to 11 percent today.

BORLAUG, THE SON of tough Norwegian American farmers, grew up on a small, hardscrabble homestead near Cresco, Iowa, during the Great Depression. He worked his way through the University of Minnesota's school of forestry, taking off semesters for stints with the Civilian Conservation Corps, where he first saw true hunger among the poor men working beside him. He never forgot how a full belly brought them back to life.

Shortly before his graduation, Borlaug attended a lecture on wheat rust entitled, "Those Shifty Little Enemies That Destroy Our Cereal Crops." The man at the podium was the renowned scientist Elvin Charles Stakeman. "Stake," as he was known, was small in stature but a lion of a lecturer, able to sway politicians and presidents of industry. He was the nation's expert on wheat rust and led the American eradication effort after the blight ravaged the wheat belt during World War I. He served on the board of the National Science Foundation, as president of the American Association for the Advancement of Science, even as an adviser to the Atomic Energy Commission. In his 40-acre lab at the University of Minnesota, he studied nearly every known disease of cereal crops in the Midwest.

"Stake's object is to develop tough new varieties of wheat and other cereals that will resist these diseases," a *Time* magazine reporter wrote in a 1944 profile. "His $300,000 laboratory at the University of Minnesota is one of the liveliest in the US, pulsing with Arrowsmithian fervor. His graduates, scattered over the earth, today are fighting fungi in Europe, Australia, China, India." A 1952 book named Stakeman as one of the 100 most important people in the world.

Stakeman was an acute judge of character. He'd noticed Borlaug not in class, but on the college wrestling team. Norman wasn't the most skilled on the squad, which included a few national champions, but he had an unbreakable determination that took him to the Big Ten semifinals in his weight class. When Borlaug asked Stakeman whether he should pursue a graduate degree in forest pathology, Stakeman convinced him to come over to field crops, where he could work on rust resistance and do some real good. Here, the farm kid found his calling. Borlaug was soon working on his doctorate in plant pathology and genetics under Stakeman. After graduation, he married his college sweetheart and accepted a lucrative job at DuPont just as the Japanese bombed Pearl Harbor. He tried to quit and enlist but was told he couldn't; his lab had been deemed critical to the war effort.

In 1943, at the request of the New York–based Rockefeller Foundation, Elvin Stakeman lead a fact-finding mission to Mexico. Land reform and rust epidemics had devastated the country's wheat production; the nation was becoming more and more dependent on US imports. Established by Standard Oil founder John D. Rockefeller in 1913, the Rockefeller Foundation was the largest philanthropic organization in the world at the time. After World War I it had spent $22 million to charter its own ships to deliver food to the starving, war-ravaged populations of eastern Europe and had initiated numerous health programs in impoverished areas of the southern United States. After a tour of rural Mexico, US Vice President Henry Wallace, a legendary plant breeder himself, had asked foundation president Raymond Fosdick if he could help Mexican farmers boost their production. Stakeman and his team rented a run-down farm in the Toluca Valley outside Mexico City, and Rockefeller's "Mexican Agricultural Program" was born.

Officially known as the Office of Special Studies (OSS) within the Mexican Ministry of Agriculture, the team included experts in corn, beans, wheat, and potatoes. Stakeman recommended his top student, George Harrar, to run the program. Many agricultural historians believe that Harrar was the true father of the green revolution. Yet his greatest strength was in picking the right people for the job. Barely a year into their work in Mexico, Harrar and Stakeman convinced Norman Borlaug to leave a secure, lucrative position in the fungicide lab of DuPont Chemical (where he worked on DDT, among other things), to come join them as their plant pathologist and wheat specialist.

The story of Borlaug in Mexico is now legendary. He arrived with little equipment, no trained staff, no decent housing. He swept the rats out of a dilapidated agricultural warehouse and slept on the floor. He was wracked by dysentery from the bad food and water. At one point he tied a rope to a heavy cultivator, threw it over his shoulder, and manually pulled it through a field to prepare an early test plot for planting. Local farmers thought he was nuts, or a government spy, and offered little help. It did not take long before they became his staunchest defenders.

Borlaug's first goal was to breed rust resistance into the long-stemmed Mexican wheat. Plant breeding in the 1940s was a painfully slow process. Breeders would carefully select varieties containing desirable traits, such as yield or milling quality, to cross with a variety that might have better resistance to rust. They'd plant the offspring and then wait six months for the results. The selection and crossing ritual continued in succeeding generations until they achieved a stable mix of the traits they were after. In most places, they could grow only one crop a year.

But Borlaug was in a hurry. He and his colleagues gathered wheat germplasm from around the world, as well as the best Mexican varieties, and then, as his former student Dr. Ronnie Coffman of Cornell says, "he just crossed the hell out of them." Using Stakeman's protocols, Borlaug made hundreds and then thousands of crosses, creating a huge genetic swamp. He then ruthlessly culled through the worthless plants to find the traits he wanted. After a few years and 5,000 crosses, he had only two that showed rust resistance. At this rate, he figured, it would take 10 years to get the wheat he wanted. Borlaug learned of an abandoned agricultural experiment station 800 miles north in the dry but irrigated Yaqui Valley of Sonora. The wheat-planting season there began just as harvest was ending in Toluca. Borlaug figured if he could harvest his best crosses from Toluca and hustle them up to plant in the Yaqui Valley, he could get two growing seasons a year and double the number of crosses. He could then take the best seeds from the Yaqui Valley and rush them back to Toluca to plant.

Harrar thought the plan absurd, and not just because of the additional cost. No roads connected the two areas in Mexico. Borlaug would have to haul his seeds to Texas, drive to Arizona, and then come back down into Sonora during the rainy season, when many of the roads and bridges were flooded. There were serious scientific questions as well. At the time most breeders, including Harrar, believed that wheat seed needed a rest period after harvest before it would be viable. Borlaug wanted to plant the seeds from his crosses immediately. More important, most wheat varieties were naturally sensitive to day length to encourage flowering. Borlaug's two locations were separated by 10 degrees of latitude and 2,200 meters of elevation and had vastly different climates, day lengths, and seasons. Breeders at that time believed that wheat varieties could grow only in the areas where they were adapted to local growing conditions.

Stakeman intervened between his former students and con-

vinced Harrar to give Borlaug's plan a try. The gamble paid off.
Borlaug discovered that by exposing his crosses to different envi-
ronments, stresses, and diseases, the wheat plants that survived
were robust, adapted easily to various climates and day lengths,
and had a wider range of disease resistance. Borlaug had invented
"shuttle breeding," a technique still used by plant breeders today.
By 1948, Borlaug's team had developed several varieties that
offered better yields and greater protection from stem rust, and by
1955 he had achieved complete control of the disease in Mexico.
In 1956, Mexico produced a million tons of wheat—four times the
harvest of 1945—and became self-sufficient in the crop.

But rust, as Borlaug famously said, never sleeps. In 1952, a new
strain of stem rust, called 15B, destroyed two-thirds of the bread
wheat harvest in North America and then hit South American
fields the following season. The commercial varieties had no resistance to it. The USDA, with Stakeman's guidance, established
the International Spring Wheat Rust Nursery program in several
countries to provide a rust-resistant gene pool. The OSS contributed heavily to the nursery, and the rust-resistant complex of genes
that Borlaug's team developed, known as $Sr2$, remains the basis of
many rust-resistant varieties planted around the world today.

The Mexican farmers, particularly those in the irrigated Yaqui
Valley, were delighted with the new wheat seeds; the more fertilizer they applied, the more they seemed to yield. But there was a
catch. The plants also grew taller with all the extra nutrients, and
that tall stalk was easily blown over by wind—what farmers call
"lodging." Lodged wheat is hard to harvest, and much of it simply
rots on the ground.

Borlaug and his colleagues scoured the global seed banks looking for stronger-stalked wheats that they could cross into their
Mexican varieties, but without much luck. Then, in late 1952, Borlaug heard that Dr. Orville Vogel at Washington State University
was experimenting with a dwarf wheat variety from Japan. The

poor, spindly wheat, known as Norin 10, was, in retrospect, one of the most valuable prizes of World War II. A plant geneticist named S. C. Salmon, assigned to General Douglas MacArthur's occupation army, had come across the short-stalked wheat in the lab of Dr. Gonjiro Inazuka, the chief wheat breeder at an agricultural experiment station in northern Japan. Salmon had collected seed samples from a number of dwarf varieties and sent them to Vogel. Vogel had made a few crosses with them, none very good. The best was a cross between Norin 10 and a US variety named Brevor. This was one of four semidwarf wheats that Vogel sent to Borlaug in Mexico.

The OSS program was a decade old the first time Borlaug crossed Norin/Brevor with his Mexican spring wheats, in 1954, and the results were terrible. Rust nearly wiped out the entire test plot, leaving only a handful of viable seeds. Many of the early crosses were sterile, or their seeds were shriveled, or they had awful milling quality. Over the next few years Borlaug and his team crossed the bad traits out and bred in the good. By the late 1950s they were getting short, stout-stemmed wheat with excellent milling quality and rust resistance. It was also fast maturing—which meant farmers could plant two crops a season in places. And because shuttle breeding made it highly adaptive, it could be sown virtually anywhere wheat was grown. Most important, it hungrily sucked up water and fertilizer and poured the extra energy into the seed head. Yields were double that of the long-stemmed wheats. Borlaug held his first field day to show farmers the new cultivars in the spring of 1961. According to one of the researchers there that day, the Yaqui Valley growers were so excited that they ran into the thick green carpet of wheat and began ripping the heads off the stalks, unable to believe their eyes.

In 1965, Mexico harvested 2.5 million tons of wheat—a 10-fold increase since Borlaug's arrival two decades earlier.

For those who don't wrest their living from the land, the significance of this feat is hard to appreciate. But consider this. It took a thousand years for farmers to increase wheat yields from half a metric ton per hectare to 2 metric tons per hectare. Thanks to Borlaug and colleagues, it took only 40 years to increase those 2 tons to 6. Borlaug became a folk hero. A rumor even started circulating around northern Mexico that Borlaug "talked" to each plant before he selected it for his crosses.

WHILE BORLAUG WAS consumed with his wheat, the Cold War was unfolding in Asia, where peasant-backed Communist revolts swept through China, Korea, and Indochina. India, the world's second-most-populous country, experienced Communist uprisings as well. Many strategists believed that hunger in Asia was aiding the Communist cause. John Kerry King, a Fulbright scholar who would later become director of the Central Intelligence Agency's Office of Political Analysis, stressed the political power of rice in a 1953 article in *Foreign Affairs*:

> In a restless Asia . . . the question of the supply of this commodity upon which life itself is based has far-reaching political implications. . . . In 1952, Communist China emerged as an exporter of rice—a dramatic change after so many years of scarcity; and the Communists have made much of the fact in their propaganda in Asia. . . . The struggle of the "East" versus the "West" in Asia is, in part, a race for production, and rice is the symbol and substance of it.

As the Cold War intensified, the Ford and Rockefeller Foundations together, encouraged by the success of Rockefeller's Mexican program, began contemplating a similar effort for Asia, where rising hunger, poverty, and populations were fomenting political unrest. "At best the world food outlook for the decades ahead is grave; at worst it is frightening," argued Ford Foundation Vice President Forrest "Frosty" Hill, concluding that the threat posed by global population growth and food supply was second only to "the possibility of all-out nuclear war."

George Harrar, now the president of the Rockefeller Foundation, met Hill quietly at the University Club in New York City, and together they hammered out a deal to build a $7 million agricultural research center in the Philippines. The International Rice Research Institute (IRRI) became what green-revolution historian Nick Cullather dubbed a "Manhattan Project for food." But unlike the infamous nuclear weapons program, it was far from secret. Built on the outskirts of Manila on the slopes of a volcano, IRRI was to be staffed by top international agricultural scientists who, like Borlaug and his colleagues in Mexico, would train a cadre of agronomists, geneticists, irrigation specialists, plant pathologists, and agricultural economists to spread new high-yielding rice throughout Asia.

The traditional method of cultivating rice was at the foundation of Asian agrarian cultures. If IRRI scientists could develop superior rice varieties, even illiterate peasant farmers might acknowledge the technical superiority of the West. More important, high yields, combined with loans for inputs like seeds, fertilizer, pesticides, and irrigation, would produce profits, which communal peasant farmers would spend on radios, bicycles, and other goods, leading them down the road to consumer capitalism instead of revolution. The same logic inspired the famous gadget-filled American kitchen that Nixon showed Khrushchev in 1959.

Using food as a political weapon has a long history. Herbert Hoover dangled food aid to eastern Europe during the 1920s to help stave off "Bolshevist insurrections," while Truman funneled food to postwar Europe under the Marshall Plan (1948–52) to help keep both famine and communism at bay. Such government

purchases boosted US farm prices and production but led to growing stockpiles when the programs ended. To get rid of the surpluses but maintain price subsidies, Congress passed Public Law
480, the Agricultural Trade Development Assistance Act of 1954,
which enabled hungry countries to get cheap US commodities on
easy 30-year credit or even for free—at the expense of US taxpayers. Signing the legislation, President Eisenhower claimed it
would "lay the basis for a permanent expansion of our exports of
agricultural products with lasting benefits to ourselves and peoples
of other lands." John F. Kennedy, never at a loss for good slogans,
renamed the program "Food for Peace" and placed two of its food
programs under the control of his new US Agency for International Development, (USAID).

When early critics of the program argued that the United States
would be creating beggar nations, Senator Hubert H. Humphrey,
from the farm state of Minnesota, replied: "To me that was good
news, because before people can do anything they have got to eat.
And if you are looking for a way to get people to lean on you and
to be dependent on you, . . . food dependence would be terrific."

Borlaug saw politics as an impediment to production; he was
often ensnared in its web. In 1954, Mexico's undersecretary of
agriculture even tried to toss him and the entire OSS group out
of the country for promoting US and Rockefeller interests above
Mexico's. The undersecretary relented only after farmers and agricultural leaders came to the program's defense.

The new International Rice Research Institute in Los Baños,
Philippines, however, had the full blessing of the Philippine government, in particular its new president and future dictator Ferdinand Marcos—who planned to take full credit for its success. He
did not have long to wait. After Borlaug's development of semidwarf wheat, IRRI's top plant breeder, Peter Jennings, and plant
physiologist, Akira Tanaka, visualized the perfect rice plant along

the same lines: short stemmed, highly adaptable, disease resistant, with great response to fertilizer. Like wheat, traditional rice varieties were more than a meter tall and tended to lodge when fed too much plant food. But unlike wheat, dwarf or semidwarf varieties of rice were rare. Jennings's first task at IRRI was to build a seed bank of rice varieties. He and IRRI geneticist T. T. Chang wrote to rice researchers in 60 countries asking for seed samples, which soon began trickling into Los Baños.

In 1960, Chang and colleague Sterling Wortman personally scoured Asia for short, stubby rice. They finally found one in India, a Taiwanese variety named Taichung Native 1 (TN1). It yielded much better than traditional tall varieties but fell far short on disease and pest resistance. The pair also found a short parent of TN1 named *dee-geo-woo-gen* (DGWG), which translated into "brown-tipped sharp-legged thing," which they thought might be the genetic source of TN1's height.

What happened next, Jennings admitted, was "just sheer luck." In late 1962, he and his colleagues planted their first crosses—38 of them. It was a ridiculously small number even by 1960s standards. It had taken Borlaug more than 8,000 crosses to find a semidwarf wheat worth planting. The IRRI team used three short Taiwanese varieties, as well as a few tall varieties that were known producers, including one called Peta from Indonesia. The first-generation hybrids were awful—6-foot-tall giants that were worse than either parent. Most were sterile. The eighth cross, between DGWG and Peta, produced only 130 seeds. But having nothing else, they harvested the seed and grew out a large second, or F_2, generation, planting 4,000–6,000 seeds from each single cross. Then they transplanted the seedlings into their test paddies.

After a month the researchers walked out to the field to check on the F_2's progress. The first plot they came to was a mess, a total jungle. The next six were equally bad. Then Jennings arrived at the eighth plot, which held the second generation of the Peta-DGWG cross, and he felt as if he'd been transported to Gregor Mendel's pea garden. Mendel, of course, was the Austrian monk whose work breeding pea plants had unlocked the secrets of heredity. When Mendel crossed purebred tall peas with purebred short peas, he got three tall pea plants to every short one. The characteristic that gave this result came to be known as a single-gene recessive trait for shortness. In just 38 crosses and two plantings, Jennings had found the elusive gene for shortness in rice.

"It was an epiphany!" Jennings told an interviewer in 2007. "I never had an experience like that in my life—before or since. There were tall plants and there were short plants, but there were no intermediate plants! The short ones were erect, darker green, and had sturdy stems and a high number of tillers. We counted the tall plants and short plants. Essentially, the ratio of tall to short was 3 to 1—obviously a single gene recessive for shortness! It may sound something like arrogance, but I contend that I knew, at that moment, the significance of this."

Jennings persuaded Bob Chandler, the director of IRRI and a veteran of Rockefeller's Mexican program, to hire Henry "Hank" Beachell, a renowned rice breeder from Texas A&M to help refine the line and breed in disease and pest resistance. Beachell had spent years irradiating rice seeds with gamma radiation in an attempt to induce dwarf mutant strains, with no luck. By 1966, less than four years after the first cross, a young Indian agronomist named S. K. De Datta had conducted fertilizer response trials to see what the new variety, now known as IR8, could do under best management practices. IR8 averaged 9.4 tons of rice per hectare, with one plot yielding more than 10 tons. A fellow researcher in Pakistan reported an astounding 11-ton harvest—*more than 10 times* the world's average rice yield at the time. Moreover, IR8 wasn't sensitive to day length and it matured in a mere 130 days, as opposed to 160 or 200 days for most rice of the day. That meant farmers in the tropics could plant not just two but sometimes even three crops in one year. De Datta and Beachell showed the results to Chandler. "The whole world will hear about this," he told them. "We're going to make history!"

The Western media dubbed IR8 "miracle rice," and it remains perhaps the most celebrated crop in the long annals of agriculture. Never mind that it was tough and chalky and broke during milling. Those qualities could be improved later. What mattered to Ferdinand Marcos and Lyndon Johnson was that it yielded like crazy and put traditional varieties to shame. Both Marcos and LBJ visited IRRI in October 1966 to see the new rice, or rather, to be seen with the new rice. Jennings remembered the day well. IRRI staff had built a platform out into a test plot of IR8 so that the two presidents wouldn't have to soil their shoes in the Philippine mud. Jennings, Beachell, and Chandler headed out to the platform with Marcos and LBJ trailing behind. As they approached the rice, Jennings, in his mid-30s at the time, heard Johnson's deep Texas drawl behind him. "Boy!"

"Sir?" Jennings replied, thinking LBJ had a question about IR8.

"Boy, move over to one side; the photographers want to take my picture."

Later, at the podium with the cameras rolling and IR8 waving in the background, Johnson gave one of his trademark vein-bulging speeches: "If we are to win our war against poverty, and against disease, and against ignorance, and against illiteracy, and against hungry stomachs, then we have got to succeed in projects like this, and you are pointing the way for all of Asia to follow!"

Within a few years, IR8 was being planted in nearly every major rice-growing nation in the world, while national breeding programs, particularly in India, were crossing it with native varieties to adapt it to local conditions. Jennings and his colleagues at

IRRI were sending thousands of small starter packs of seed and fertilizer—enough to plant 1 hectare—to extension agents all over Asia for immediate distribution to farmers—bypassing the typical government channels. More than 2,500 Filipino farmers traveled to the IRRI campus by bus, by bike, and on foot to get their free seeds. Rice yields quadrupled from Pakistan to Colombia. IRRI quickly rolled out better versions (IR5 and IR20) with more disease resistance and better quality, but the green revolution in rice was ignited by IR8. Even though the population of East and Southeast Asia nearly doubled between 1961 and 1991, rice production easily outpaced population growth, increasing from 52 to 125 million metric tons over the same period.

THE VARIETAL PROVED useful in another war. Johnson left IRRI immediately after his speech to go visit Vietnam, where US forces were using IR8 as a propaganda tool. The South Vietnamese government had christened it *nong*, "rice of the agricultural gods," and North Vietnamese farmers were at one point inundated by thousands of leaflets proclaiming that South Vietnam was experiencing a "rice revolution" and that all Vietnamese would share in the "miracle rice" after the peace.

Vietnamese farmers, being of a practical bent, called it *luo Honda*, or "Honda rice," because one good crop would buy a new motorbike. But it didn't have the political impact that US army and intelligence agencies had hoped for. Once the Vietcong saw IR8's potential, they immediately smuggled it into North Vietnam and used it to bolster their own image as great providers of modern technology and food for the common man. The fact that it was the creation of research funded by two of the world's greatest capitalists was lost in translation. Oddly enough, Vietnam is the only place in the world today that still grows IR8, the original miracle rice.

At the same time IRRI was rolling out IR8, Norman Borlaug was taking his superwheats on the road. During 1965 and 1966, back-to-back droughts were wreaking havoc on India, which had lost much of its fertile wheat lands to Pakistan during partition. The United States was the world's only country with food reserves available for export at the time, and it shipped much of its surplus to India. The country was growing increasingly dependent on PL-480 wheat; as much as 10 million tons was imported in 1966 alone, leading some to comment on India's "ship to mouth" existence. A few "neo-Malthusian" academics in the West went so far as to suggest food-aid triage—essentially cutting off aid to countries deemed to be growing far beyond their carrying capacity and letting Mother Nature take her ruthless course.

But for President Johnson, America's bounty was a powerful tool for political leverage. After Kennedy's assassination, LBJ placed new restrictions on US food aid. Recipient nations now had to shift their development efforts from industry to agriculture, launch or expand family-planning programs to bring population growth under control, and open their doors to US investors—a nod to US corporations that wanted to build fertilizer and pesticide plants on the subcontinent. (Union Carbide built one such plant in 1969 to produce the carbaryl insecticide Sevin. The plant exploded 15 years later, killing an estimated 18,000 people in the surrounding town of Bhopal in one of the worst industrial accidents ever recorded.) Indian Prime Minister Indira Gandhi was an outspoken critic of the Vietnam War. As a result, Johnson reduced US food aid to India to a trickle—just enough to avert widespread famine.

Borlaug had visited India earlier in 1963 at the invitation of a young plant breeder named M. S. Swaminathan at the Indian Agricultural Research Institute (IARI) in New Delhi. Swaminathan had seen Borlaug's semidwarf wheats at the USDA's International Spring Wheat Rust Nursery and thought they might be suitable for

India, particularly the rich, irrigated soils of the Punjab, Haryana, and western Uttar Pradesh. Borlaug had already sent some of his seeds to researchers in Pakistan's Punjab, and they were thriving. During their travels through the wheat fields of India, the two men hit it off. "It was a wonderful experience traveling with him since I found him to be not only a brilliant scientist, but a humanitarian to the core," Swaminathan later wrote. "I referred to Dr. Borlaug as the Albert Schweitzer of Agriculture," after the legendary medical missionary of Africa.

After his visit, Borlaug sent 100 kilograms of seed from four semidwarf Mexican varieties to Swaminathan, and together they devised a five-year plan (1963–68) to bring the wheat revolution to India. The two tried to import large quantities of Mexican wheat seed to the subcontinent, but they were stymied by India's big parastatal seed companies, red tape, and political opposition. To achieve their yield potential, the new seeds required new fertilizer factories, along with tremendous government investments in irrigation, agricultural extension workers, and farm credit for small farmers. Not until 1966, with famine looming, did Gandhi's government finally relent and approve Swaminathan's proposal to import 18,000 tons of Borlaug's seeds from Mexico to set up demonstration plots. Swaminathan insisted that these test plots be planted in the fields of poor farmers—not wealthy ones—to prove that the new seeds could be grown by anyone. The farmers were supplied with all the necessary inputs and given technical advice, but they did the work themselves. The small farmers quadrupled their yields. Swaminathan then set up seed villages to start rapidly propagating the Mexican seeds. A year later, Pakistan imported 42,000 tons of the Mexican seeds—the largest seed import in history.

The story of the first major shipment of seed from Mexico is almost as wild as that of IR8 in the Philippines. The 35 trucks carrying the seed from Mexico's Yaqui Valley to the port of Los Angeles were first held up by customs officials at the border. They finally arrived in mid-August 1965, just as the Watts race riots were erupting in Los Angeles, causing the National Guard to shut down the highway to the port. The trucks made it to the ship just a few hours before its departure, and then the shipping agents found a problem with India's letter of credit to pay for the freight. Borlaug, who had been working the phones for nearly 48 hours straight, signed the letter of credit himself, without even asking for approval from his superiors. His boss, George Harrar, seriously considered firing Borlaug during this period because his expense accounts and paperwork were always late. Borlaug also threatened to quit. "What do you want, paper or bread?" he'd ask Harrar.

With his seeds now on their way to India, Borlaug finally turned in for some much needed sleep. He awoke the next morning to the news that India and Pakistan were at war. Borlaug had to arrange for the wheat to be off-loaded in Singapore, divided among two different vessels, and sent separately to each country.

Borlaug got his reward at the end of the growing season. After the seeds were planted in the fertile soils of the ancient Indus Valley, the yields in both countries obliterated all previous production records. Pakistan produced 4.6 million tons of wheat in 1965—the most it had ever grown. With the introduction of Mexican wheat, its production grew to 6.7 million tons in 1968, 7.2 million in 1969, and 8.4 million in 1970. The nation, at the time known as West Pakistan, became self-sufficient in wheat in 1968.

India's yield increases were just as dramatic. From a record 12.3-million-ton wheat crop in 1964–65, India's wheat production rose to 16.5 million tons in 1968, 18.7 million in 1969, and 20 million in 1970. In 1968, Stanford biologist Paul Ehrlich argued in *The Population Bomb* that it was sheer "fantasy" that India could ever feed itself. Yet that year, when the first major crop of Mex-

ican wheat ripened on nearly a million hectares of Indian farmland, there were not enough people to harvest it all, not enough
machines to thresh it, and not enough warehouses to store it. The
Indian government closed village schools early and turned them
into granaries. The combination of Borlaug's wheat and IRRI's rice
made India self-sufficient in food grains by 1974. By 1991, wheat
production in South Asia had nearly quadrupled.

In the spring of 1968—with bumper crops of the new wheat
and rice being sown, reaped, and heavily promoted by the United
States all over Asia and Latin America—William S. Gaud, the
administrator of USAID, tried to describe the new kind of agriculture that was sweeping the globe. "These [record yields] and other
developments in the field of agriculture contain the makings of a
new revolution," Gaud told the Society for International Development, whose members had gathered in a Washington, DC, hotel
a few blocks from the White House. "It is not a violent Red Revolution like that of the Soviets, nor is it a White Revolution like
that of the Shah of Iran. I call it the Green Revolution. This new
revolution can be as significant and as beneficial to mankind as the
industrial revolution of a century and a half ago."

Fifty years after the green revolution transformed world agriculture, its legacy is still hotly debated. The benefits were substantial, especially in Asia. The number of calories consumed per
capita increased by nearly 30 percent, while the price of wheat and
rice dropped and stayed low. Even with lower prices the higher
yields stimulated the rural economy. Real per capita incomes
nearly doubled between 1970 and 1995, while the number of people in poverty was cut in half. Oddly enough, the monocultures
that produced cheap staples enabled people to afford more diversified diets, so they began eating more fruits, vegetables, oils, and
livestock products. Noted economist Jeffrey Sachs of Columbia
University goes so far as to credit the green revolution for Asia's
emergence as an offshore manufacturing powerhouse.

Incredibly, all that extra food was produced on nearly the same amount of land. Borlaug and his supporters long argued that the high yields of the green revolution saved hundreds of millions of hectares of wilderness from the plow that would have otherwise been needed to produce food for hungry people. (Recent modeling puts the figure at a more modest, but still impressive, 20 to 30 million hectares spared.) Borlaug himself is credited with saving millions of lives from starvation.

Critics of the green revolution focus on the enormous cost it levied on both people and the planet. Biologist and writer Rachel Carson was among the first to speak out against the threat of pesticides, one of the innocuously labeled "inputs" needed to protect the yields of densely planted fields. Pesticide use tripled between 1960 and 1990, leading to increased farmworker poisonings, wildlife deaths—including both birds and bees—and residues in food. The global use of fertilizer increased more than eightfold between 1960 and 2000, contaminating rivers, lakes, ground-water, and estuaries. Harmful algal blooms and dead zones fueled by excessive nitrogen and phosphorus are now commonplace in coastal estuaries around the globe. The annual dead zone off the Mississippi River covers about 5,000 square miles of the Gulf of Mexico—an area the size of Connecticut where no fish can live.

The green revolution caused social problems as well. The bulk of the benefits flowed to larger landowners, while the increased use of machinery reduced the need for agricultural labor, leading to an influx of unskilled peasants into the cities—just as the Enclosure Acts and threshing machines had done in England a century before.

Even Borlaug's biggest fan and collaborator in India, Dr. M. S. Swaminathan, quickly grew concerned about the new farm mentality that put profits over soil health. In his 1968 speech to the Indian Science Congress, Swaminathan seemed to channel the "Ghost of Agriculture Yet to Come," warning his colleagues of the

consequences should the green revolution turn into the "greed revolution." Intensive cultivation of land, Swaminathan said, without
conservation of soil fertility and structure would lead to deserts;
irrigation without drainage or unscientific tapping of groundwater
would lead to salinization and rapid exhaustion of the resource;
indiscriminate use of pesticides could lead to cancers and other
diseases; and the rapid replacement of locally adapted varieties
with one or two high-yielding strains could lead to disease epidemics like those that had triggered the Irish potato blight or the
Bengal famine.

> The initiation of exploitive agriculture without a proper under-
> standing of the various consequences of every one of the
> changes introduced into traditional agriculture, and without
> first building up a proper scientific and training base to sus-
> tain it, may only lead us, in the long run, into an era of agri-
> cultural disaster rather than one of agricultural prosperity.

NO ONE WAS more keenly aware of the limitations of the green
revolution than Borlaug himself. Borlaug was a tough but humble man. He never registered a single patent on his seeds—whose
market value has been estimated in the billions—but instead gave
them away freely. He accepted the Nobel Peace Prize on behalf
of all the scientists and farmers who brought the green revolution
to fruition. He remains the only agricultural scientist ever to win
the award.

Borlaug's Nobel lecture in December 1970, like Malthus's lifelong compassion for the poor, is long forgotten. But it might
surprise both his modern critics and boosters from agricultural
corporations that funded his later work. In it, Borlaug spoke of
food being a moral right, the first component of social justice.

He spoke of the need for international granaries to prevent famine, the need to diversify crops and plant gene pools, the need for the preservation and free flow of seeds to researchers around the world, and the importance of training local agricultural scientists, whose primary goal was to serve not corporations, universities, or governments, but all farmers and thus all humanity. A decade before Amartya Sen's 1981 *Poverty and Famines*, Borlaug made a similar plea: it's not enough just to increase food production; we have to increase the purchasing power of the "vast underprivileged masses" to improve their access to the new agricultural bounty.

Most striking, however, Borlaug warned that the green revolution, even if fully implemented, would give humanity sufficient food for only 30 years. "The frightening power of human reproduction must also be curbed," Borlaug warned. "Otherwise the success of the green revolution will be ephemeral only. . . . Most people still fail to comprehend the magnitude and menace of the 'Population Monster.' . . . There can be no permanent progress in the battle against hunger until the agencies that fight for increased food production and those that fight for population control unite in a common effort. Fighting alone, they may win temporary skirmishes, but united they can win a decisive and lasting victory to provide food and other amenities of a progressive civilization for the benefit of all mankind."

It is perhaps the ultimate irony that the father of the green revolution, the man most responsible for destroying the credibility of the neo-Malthusians of his day, was the most ardent Malthusian of them all.

With bare arms and a backpack sprayer, a farmworker applies pesticide in a cotton field near Jajjal village, Punjab, India.

CHAPTER 4

The Plight of the Punjab

In modern times mechanization has changed the age-old agricultural practices. But there is no replacement for efficient peasants.—*Poster hanging in the National Museum of Agricultural Science, New Delhi, India, October 2008*

t's a long, hot, bumpy road from Delhi to Chandigarh, the capital of the Indian state of Punjab. Blaring horns, car-eating potholes, garish buses, and lifesaving swerves make the drive feel like an amusement park ride spinning dangerously out of control. I arrived during the *kharif*, or fall harvest season, when the paddy—as the rice crop is called here—was being "lifted" and carried to the local grain markets for threshing and sale. An intense dry heat baked the flat gray fields and mud-brick villages, the horizon broken only by the occasional coal-fired

power plant with its giant smokestacks darkening the sky. Smoke
billowed from the squat chimneys of brick kilns that dotted the
countryside, while black storm clouds of soot gathered above harvested rice fields set ablaze to clear the land for planting wheat.
Dust roiled up in chalky waves from the wheels of a million old
buses, rounded Ambassadors, square Mahindra jeeps, wildly
painted lorries, and an endless procession of motorcycles driven
by young men wearing mirror shades and bandannas. Often they
carried women riding sidesaddle behind them in flowing yellow
and orange saris, one hand grasping the driver, the other clutching a tiny child. Once, we passed a gated development of new,
whitewashed apartment buildings rising from the fields like a
mirage in the desert. The billboard out front beckoned wealthy
Punjabis to "Come Live the Canadian Lifestyle!"

More often we passed farmhouses with water tanks shaped like
giant soccer balls or birds; or, in the dusk of the evening, gleam-
ing wedding palaces like mini–Taj Mahals draped in waterfalls of
colored lights with thin, quiet men drinking out of paper cups outside. Yellow cur dogs roamed the ditch banks dragging swollen
teats because, aside from being the harvest season and the festival
season and the wedding season, it also happened to be puppy season in the fecund land of Punjab, the ancient alluvial fan of the
Indo-Gangetic Plain.

Life is supposed to be good here. Though they inhabit one of
the smallest states in a very large nation, Punjabis enjoy one of
India's highest per capita incomes and are not shy about showing
it. Punjab has been the backdrop of countless Bollywood films,
filled with lush green fields owned by opulent Punjabi farmers and
their stereotypically cocky, jeep-driving 20-something children.

Much of that swagger—and much of India's phenomenal recent
economic boom—is rooted in the green revolution. When Norman Borlaug and M. S. Swaminathan introduced the new high-

yielding varieties of wheat and rice to India, they focused their initial efforts in Punjab, a traditional wheat-growing region on the border with Pakistan. The result was a staggering surplus of grain. In 1960–61, Punjab farmers harvested about 3 million tons of wheat and oilseeds—not even enough to feed all Punjabis. In 2004–5, they harvested more than 25 million tons of grain, virtually all of it high-yielding varieties of wheat and rice—enough to feed Punjab and much of India as well.

Punjab was singled out for the new kind of agriculture because it had the plentiful water that Borlaug's high-yielding varieties needed. The word "Punjab" comes from the Persian root *panj* ("five") and *aab* ("water"). It is the land of five rivers: the Jhelum, Chenab, Ravi, Beas, and Sutlej gather meltwater from the snowy Himalaya and Pir Panjal mountains to the north and east and carry it across fertile but arid floodplains southwest toward Pakistan, where they feed the Indus.

This is the land of the ancient Harappans, some of the world's earliest irrigators, whose agricultural prowess 4,500 years ago sustained a civilization larger than Egypt and Mesopotamia combined. They were the first farmers to cultivate cotton, gin it, spin it, and weave it into cloth, which they traded to the neighboring Mesopotamians. Though invading Aryan herders destroyed much of the Harappans' hydraulic works, the British eventually recognized the agricultural potential of the region, and during the late 1800s they sought to create a "model agricultural province" amid a network of irrigation canals. They settled these new canal colonies with loyal Hindu and Sikh farmers who had formed the backbone of the Indian Army for decades. With access to water, these farmers turned 6 million acres of desert into what is now the agricultural dynamo of India and Pakistan.

When the Indian government saw the potential of Borlaug's wheat varieties in 1968, it spent millions on agricultural development. Over the next decade, India's leaders funneled the money into the five states that they thought had the greatest potential: Andhra Pradesh, Tamil Nadu, Gujarat, Rajasthan, and Punjab. The highly literate Punjabi farmers quickly adopted the new package of seeds, fertilizers, and pesticides—especially the larger, wealthier farmers, who reaped the greatest rewards. But even the smaller farmers prospered. The government subsidized cooperative banks, making farm credit incredibly cheap. Small Mahindra, Farmtrac, and Eicher tractors were soon crawling over the Punjab, becoming status symbols among even the smallest landowners, who either ignored or never heard the warnings of economists from Punjab Agricultural University that it would take at least 15 acres of land to make such a machinery investment pay off. Most had far less.

It didn't matter. The government was buying every grain of wheat and rice at a decent support price, and subsidizing nitrogen fertilizer as well. Only about a quarter of Punjab was serviced by the old British canals. So with help from the World Bank, the primary funder of national irrigation schemes, the Indian government also subsidized the installation of tube wells—small-diameter steel pipes attached to electric pumps—so that farmers could tap the shallow and abundant groundwater. The government helped farmers install 200,000 tube wells in just two years in the early 1970s and provided the electricity to run the pumps for free. Today, nearly 1.3 million tube wells pierce Punjab's aquifer in the central and northern parts of the state.

With a steady stream of water, subsidized fertilizer, and high support prices, farmers were virtually guaranteed a good return on their harvest. In a few years Punjabi farmers went from growing more than 200 different crops to growing just 3: wheat and rice for grain, and cotton as a cash crop. Rice farming was new to Punjabis, who traditionally ate wheat. But with high yields and a guaranteed price, the thirsty crop quickly replaced traditional pulses, oilseeds, mustards, and millets in the rotation of Punjabi farms.

"Punjab state, like the rest of India, was in deficit in food grains at Partition," said Dr. P. S. Rangi, a longtime economics professor at Punjab Agricultural University, and now a consultant for the Punjab State Farmers Union. "Yet in 20 years these farmers were able to make India self-sufficient in food grains. The [annual] growth rates have been very high, as much as 11 percent. There is no comparable example in the history of the world where the growth rates have been so high for 20 years. In this way the states of Punjab and Haryana have become the food basket for the country."

Rangi, a tall distinguished Sikh, sat at a simple desk in a corner office in the farmers union building in the modern city of Chandigarh, an open concrete structure that smelled vaguely of Sevin dust. He ruffled through a stack of papers until he found the one he sought, and then continued what would soon become a less happy tale. "In the early period of the green revolution, 1965 to 1985, farmers were getting a very good price. The cost of inputs was low, total production quite high. Both public and private investment in the farm sector was high. Net returns were quite high. Farmers made quite a big profit from this enterprise. However, during the late 1980s and early 1990s, the price of inputs increased at a higher rate than the base rates of commodities. Consequently, net returns declined, so farmers began feeling the pinch."

Prices weren't the only problem for farmers. The once fertile alluvial soils and the abundant, easily accessible river and groundwater began to falter under the endless wheat-rice or wheat-cotton rotations. Before the green revolution, a third of the Punjab's cropland lay fallow every year. Afterward, cropping intensity (the number of crops harvested during a given year) rose to 190 percent—nearly two crops a year—one of the highest cultivation rates on Earth. The intensive irrigation required for this level of farming caused the water table to plummet in areas served by tube wells, while areas serviced by canals have since been plagued with waterlogging and salt buildup that have made more than 80,000 hectares barren.

The green revolution left a heavy chemical legacy as well. Punjabi farmers apply more fertilizer and pesticides per hectare than do any other farmers in India. Annual fertilizer use increased eightfold from 1970 to 2006, when it reached nearly 1.7 million tons, while pesticide use has more than doubled since 1980. Most of the pesticides (75 percent) are sprayed on the expensive cotton crop, which is grown in rotation with wheat in a wide swath of southern Punjab known as the Malwa Plains. Since the late 1980s, the rising cost of fertilizers, pesticides, and labor, along with the cost of deepening tube wells and installing expensive submersible pumps to handle the greater depths, has left the average Punjabi farmer deeply in debt. The average amount owed is about 41,576 rupees—$800. It doesn't sound like a lot, but for the 63 percent of Punjabi farmers who cultivate 4 acres or less, it amounts to four years' worth of income. It's also four times the average debt owed by farmers in other parts of India.

As their production costs and debts grew, Punjabi farmers began to be squeezed in an economic vice. In the early 1990s, the phenomenal increase in yield that they'd enjoyed for two decades began to slow. Instead of the blistering 11 percent yearly growth in the 1970s and 1980s, the average annual rate of growth in wheat yield from 1990 to 2000 fell to 2 percent, while the rice yield grew just 1.3 percent yearly. In the last decade, wheat yield growth has fallen to less than 1 percent, while rice production has flatlined.

India's annual population growth rate is falling as well, but the country is still growing faster than its agricultural production. From 2001 to 2011, India added 181 million people—nearly equal to the entire population of Brazil, the world's fifth-most-populous country. Though its urban middle class now tops 160 million and is the fastest-growing segment of the Indian population, more than 60 percent of Indians still make their living from agriculture, and 80 percent live on less than two dollars a day.

Most of India's poor live in rural villages like Chottian, a dusty brown collection of mud houses in the Sangrur district of southern Punjab. Kartaro Kaur is one of them. Her family once grew a variety of food crops, but after the green revolution they grew only cotton in the summer and wheat in the winter. Her eldest son, Dulla Singh, borrowed money to buy a tractor and to pay for his sister's marriage. Then the American bollworm hit the Punjab with a vengeance. Farmers who had once sprayed their cotton for insects 10 or 12 times a season were soon spraying it 20–30 times with some of the most toxic pesticides on the planet. Singh borrowed money for better seeds and chemicals, and still the crop failed three years in a row. Then one day in 1988, his family found his lifeless body hanging from a tree near the field. He was 22 years old.

Dulla's younger brother Bhatti took over the farm after that, and things got a little better. He sold the tractor to pay off the debt. Then he, too, paid for the wedding of a sister and eventually bought another tractor—not because he needed one, Kaur said, but simply because he wanted one. For 16 years they could not break even. And one day in the mid-1990s, they found Bhatti hanging from a tree as well.

When I spoke with her, Kaur was 70 years old and wore a white head scarf, her face as lined as the surrounding fields. She sat on the rope bed of her simple mud-brick home with her two widowed daughters-in-law and her grandsons by her side. She told her story without emotion. Only when I asked her what her life had been like when she was young did her voice begin to waver.

"When I was young we grew mustard and chickpeas," she said. "We made more profit because the costs were much less. Those were such prosperous times forty years ago. I had so much gold then. Cotton has ruined everything."

"Why not plant other crops?" I asked.

"You can't plant the old crops, because there are so many insects that attack them," she replied. "It is very difficult."

Dusk had fallen while we spoke, and her grandsons opened their cell phones to provide light so that I could continue to take notes. I asked Kaur if she wanted them to be farmers. "No," she replied firmly. "They should get office jobs."

Kaur's sons were not alone, but rather part of a wave of farmer suicides that has shaken rural India since the late 1980s—about the same time the green revolution began to lose its glow. Exact numbers are hard to verify. Suicide is a crime in India, requiring a police investigation and a postmortem report that the victim's family must pay for. It also carries a significant social stigma, so suicides are often simply recorded as "sudden" deaths. But a recent investigation by Indian journalist Palagummi Sainath, the rural-affairs editor of the English-language paper the *Hindu*, reported that according to the National Crime Records Bureau—the official source of suicide figures in India—nearly 257,000 Indian farmers took their own lives over the 16 years between 1995 and 2010. That means that, on average, an Indian farmer committed suicide every 32 minutes during the last decade and a half. Despite much press attention, the problem seems to be getting worse. The number of suicides in the last eight years was actually higher than in the previous eight. Even a $15 billion loan waiver for India's farmers in 2008 didn't stanch the deaths.

"The first thing to go is the wife's jewelry," says Inderjit Singh Jaijee, a human rights lawyer and former member of Punjab's legislative assembly who has been pressuring the government for an accurate account of rural suicides in Punjab for years. "Then the land is pledged. They lose 2 acres this year, 3 acres next year, until they lose all 15 acres. Then they become agricultural laborers. It's a great shame to mortgage the land. A great loss of status and respect. [The farmer] goes from being called 'sir,' to being called 'hey you!' So he commits suicide." The most common method is by drinking pesticides—the very chemicals that purport to protect the farmers' livelihoods.

In the early 2000s, news reports began to surface of another disturbing phenomenon in rural villages in the cotton belt. The jet-black hair of numerous Punjabi schoolchildren, some as young as 10 years old, had turned gray—a rare condition most often associated with vitamin B_{12} deficiency or thyroid problems. Moreover, the number of cancers among middle-aged villagers, particularly women, seemed to be on the rise. Ailing patients began traveling from their rural villages to the southern Punjabi city of Bathinda to catch the train to the Acharya Tulsi Regional Cancer Treatment and Research Institute in Bikaner, a city in the neighboring state of Rajasthan. By early 2010, the *Hindustan Times*, one of the major Punjabi papers, reported that some 70 patients a day were boarding train number 339 for the eight-hour ride to Bikaner. The train is now widely known as the "Cancer Express."

Many blamed the high volume of pesticides sprayed in Punjab. Scientists and health officials have long known that pesticides, particularly insecticides that are designed to kill creatures with nervous systems and immune systems not terribly unlike our own, can harm human health. Some of the earliest and most persistent chemical pesticides, organochlorines like DDT and Lindane have been linked to soft-tissue sarcomas, non-Hodgkins lymphoma, and leukemia, as well as lung and breast cancers. Somewhat newer organophosphate pesticides, which are more acutely toxic but break down faster in the environment, have also been linked to non-Hodgkins lymphoma and leukemia.

Organophosphates are neurotoxins that kill insects by suppressing acetylcholinesterase, an enzyme critical to the functioning of bugs' nervous systems—as well as ours. It serves as the stop signal in the nerve's Morse code, and without it the system goes haywire, leading to convulsions, hyperventilation, vomiting, diarrhea, and ultimately death if exposure is left untreated. The sarin nerve gas that killed 12 people and injured more than 5,000 during the 1995 terrorist attack on the Tokyo subway system was one of the first organophosphate pesticides invented, and one of the most lethal, developed in Germany in 1938. Even the newer, less toxic organophosphate pesticides used in the Punjab and around the world can weaken human immune systems, hampering the T cells that attack tumors, and making people more susceptible to disease.

The link between pesticides and public health is a particular concern in the Punjab, where a number of studies have found high levels of DDT and Lindane in Punjabis' water, food, and blood since the green revolution began. A study conducted as early as 1974 found that of 140 wheat flour samples taken from mills across the state, 124 contained DDT and 116 contained Lindane. In 1980 researchers analyzed the breast milk of 75 new mothers in Punjab and found both pesticides in every sample. The average amount of DDT in the samples amounted to 18 times the safe consumption level recommended by the World Health Organization (WHO), while the Lindane amounts were higher than in any other country besides Japan.

An eight-year-long study in the 1990s found that some 70 percent of vegetables tested from markets around the state were contaminated with various insecticides, and 27 percent tested above the maximum residue limits, suggesting that farmers were not waiting the recommended 10 days after spraying to harvest their crops. By 2002, studies showed that the amount of DDT in Punjabi foods had fallen, but the amounts of organophosphate and carbamate pesticides were on the rise. Nearly all of the rice sampled (97 percent) and more than half the milk samples (53 percent) from 1996 to 2001 were still contaminated with Lindane, while

85 percent of the tested fruits and 71 percent of vegetables were contaminated with other insecticides.

Even though several of the pesticides used in Punjab are known or suspected carcinogens, proving that they cause high cancer rates in a specific area is difficult, especially when so many other factors, like smoking, alcohol use, or air pollution, also increase the risks of the disease. But in 2008 the Punjab Pollution Control Board sponsored a large study that surveyed more than 180,000 rural Punjabis to see who had contracted or died of the disease over the previous decade. Roughly half of the villagers lived in a region that grew mostly wheat and cotton, while the other half lived and farmed in an area known for wheat and rice. Those who farmed cotton relied mostly on canal water, while those who grew rice drank groundwater.

The results were striking: the confirmed cancer cases were almost double among the cotton-farming families, who used far more pesticides. Women in the cotton belt were by far the hardest hit, with the most common cancers attacking the reproductive system—cancers of the breast, cervix, uterus, and ovaries. While the levels of pesticides in drinking water, vegetables, and blood were also higher in the cotton-growing villages, so were other factors, such as heavy metals (arsenic for instance) in the water, alcohol use, and smoking. Residents in the cotton belt were also found to be using empty pesticide containers to store food, and spraying their cotton crop 30–35 times a season, often without using protective clothing, as compared to the 10–12 applications recommended by Punjab Agricultural University.

You don't have to walk far in any Malwa village to see the pain behind the statistics. In Bhuttiwala, a village in the Muktsar district, I met with Jagsir Singh, a 44-year-old Sikh with a blue turban and long salt-and-pepper beard who was the former sarpanch, the chairperson of the elected village council. While sarpanch, it had been his responsibility to fix the roads, see to the schools, care for the sanitation facilities and drinking-water sources, and record births, deaths, and marriages in the village register.

"Over the last four years we've had 49 deaths due to cancer," said Singh. "Most of them were young people. The water is not good. It is poisonous, contaminated water. Yet people still drink it." That number amounted to four times the annual cancer death rate found in the Punjab Pollution Control Board study.

"Do you think the green revolution was good or bad?" I asked.

"It was initially a great success for the nation and Punjab," Singh said. "Punjab still gives great food grains to the country. But the conditions for the farmers are deteriorating. They use more and more fertilizers and pesticides year after year. They have exploited these soils that used to be the most fertile in this country. They really feel a difference in the soil from when we used to use manures as fertilizer. If you tasted the soil earlier it tasted normal. Now it tastes salty."

We walked together down the dirt streets of the village, past gray walls, yellow mud-brick houses, and conical piles of dung used for cooking fires. He entered a gate and introduced me to Amarjid Kaur, a demure woman in bare feet, wearing a blue sari with red flowers embroidered on the hem. She was 40 years old and had been diagnosed with breast cancer 10 months earlier. Doctors had removed her breast, but she was still on medication. She had no family history of the disease, and though she'd been married for 22 years, she had been unable to conceive a child. Her family had already spent 100,000 rupees ($1,900) on her operation and will have to spend another 100,000 rupees on her treatment—each sum about the total average annual household income for a small farmer in the Punjab.

We walked farther down the dusty path until we came to the home of Tej Kaur, a woman of 50, who wore a light-brown scarf that draped low down her sari to cover her chest. Despite Singh's protests, she immediately knelt into the dirt to touch his feet—an ancient ritual of respect. After rising from the dust, she said she had also had a breast removed two years earlier, but that was not what pained her. She had lost her 7-year-old grandson to leukemia six years before, and the family had spent most of its savings on his treatment in Chandigarh.

I asked her husband, 70-year-old Billu Singh, if the family ever drank out of pesticide jars. He nodded. "Generally we wash the pesticide containers and use them to carry water or make drinking glasses," he said. "Nobody has told us not to do this."

Dozens of people in several Malwa villages offered similar tales of woe. Most were middle-aged women suffering from breast or ovarian cancer, but I met a few men as well. Sima Singh was a powerful looking man of 45 in a dirty green turban. A year earlier he had noticed a small growth on the side of his nose. The tumor had grown to the size of a grapefruit and had taken over the entire right side of his face. His right eye peered out from its grotesquely swollen socket as if he were the Indian Elephant Man. I asked him what he did for a living. "I work in the fields and used to spray pesticides on cotton and paddy," he replied.

Then there was Jagdeve Singh. The 14-year-old boy lived in Jajjal village in the heart of Malwa, one of the villages included in the study sponsored by the Pollution Control Board. Jagdeve's father invited me into their small mud-brick house as the sun went down, the only light coming from the flickers of a small TV showing *SpongeBob SquarePants* dubbed in Punjabi. Jagdeve was slumped over at an odd angle in a wheelchair in front of the set, unable to hold himself upright. His father, 35-year-old Bhola Singh, said they had first noticed his condition when he was two years old. He could not walk like the other children his age. He kept falling down. The doctors told Bhola that Jagdeve had spinal deterioration, and that he would not live past 20. But they did not say the cause. Several recent studies, however, have documented chromosomal damage in Indian and Pakistani cotton farmers routinely exposed to pesticides.

Bhola had no history of the disease in his family. He and his uncle Gurdeep Singh farmed 2 acres of cotton. They said no one had warned them of the dangers of pesticides until the previous two or three years. I asked them if they ever wore protective clothing. "No one wears protective clothing. We wear the same clothes we wear now," Gurdeep said, pointing to the loose cotton shirts that fell to their knees, their baggy cotton pants, their bare feet protected only by sandals. "No one has told us to do so. It cannot be done without spraying," he adds, "because of the bollworms."

The pesticides that the Malwa farmers use—monocrotophos, chlorpyrifos, and cypermethrin—are classified as restricted-use pesticides in the United States. Monocrotophos is so toxic that it was actually banned by the EPA in 1988. In North Carolina, no one can buy or apply such chemicals to crops unless they have a pesticide applicator's license issued by the state, or are personally supervised by someone with a license. To get the license, farmers have to take a written 50-question multiple-choice test and get 70 percent of the questions correct. It isn't easy. (Sample question: A sprayer is calibrated to apply 10 gallons per acre at a pressure of 20 psi. What pressure would be required to increase the output to 20 gallons per acre without changing the speed of travel or nozzle size?) I took the test as part of a weed science lab at North Carolina State and barely passed.

If you're spraying ag chemicals in North Carolina, the regulations require you to wear long pants, a long-sleeved shirt, chemical-resistant boots, chemical-resistant gloves, a wide-brimmed hard hat, eye goggles, and a chemical-resistant apron. In India, it's a different story. Many of the poorer Punjabi farmers are barely literate. Labels on the pesticide jars are often in other languages.

No farmer or laborer I met in Malwa used any safety equipment; they simply walked through the fields in sandals with cheap, leaky backpack sprayers. When I asked officials at Punjab Agricultural University about this lack of basic pesticide education, they said they run frequent pesticide safety seminars in the villages. If so, the message isn't getting through.

Instead, the Punjabi government simply blamed the cancer outbreak on polluted drinking water. But instead of tackling the problem at its source by reducing the chemical runoff from the fields, they began building small municipal reverse-osmosis (RO) water treatment plants in 45 of the worst-hit villages in the Malwa district to filter the contaminated water. The first was built in Kaouni village in 2008, in no small part because the finance minister happened to be from there. The plant displays an impressive array of stainless steel filters and drums in a small cinder-block building next to an evil-smelling village lagoon, but its source water came from a 500-foot-deep well. The water was then pumped at high pressure through increasingly smaller membranes and activated carbon filters to weed out most contaminants (though certainly not all). Villagers paid 60 rupees a month for 20 liters of filtered water each day, and had the option of buying more. The bottled water was popular with the frightened villagers, who no longer trusted the water from the pumps and canal. Even the poorest villagers I interviewed now got their water from the RO plant.

Jagsir Singh, the former leader of Bhuttiwala village, was proud of their new water system and thought it would help reduce the cancers. But the water in the canals and aquifer remain as contaminated as ever. "Nobody has suggested we use less fertilizer or pesticides," he said, which is not surprising. The fertilizer and pesticide industries are political powerhouses in India, and the soon to be most populous nation in the world is highly dependent on the grain yields and cotton income they help provide.

Before I left Malwa, I stopped by the home of Master Jarnail

Singh, a 65-year-old school principal who in 2002 had brought the attention of the press to the number of cancers in the district when he noticed so many of his students falling ill. He believes the epidemic started in 1980 when the American bollworm first appeared and began devastating their crops, leading to heavier and heavier pesticide applications.

"The green revolution has brought us only downfall," Singh said. "It has ruined our soil, our environment, our water table. It used to be we had fairs in the village when people would come together and have fun. Now we gather in medical centers. Financially we have had a downfall as well. Every home here used to have 100 grams of gold. It was a source of great pride. Now, no one has even 20 grams. They've sold their jewelry to pay for fertilizer or irrigation, and they hardly make any profit. The central government has sacrificed the people of Punjab for grain."

The solution, Singh believes, is what he calls "natural farming." He is one of a growing number of farmers in the Punjab who are now planting organic wheat. He was hoping to harvest 30 quintals (111 bushels) of wheat that year on his small plot. "There is no future in the green revolution." Singh said. "But if we go to natural farming there is a lot of potential because you are financially sound, your health will be okay, and you will be prosperous, because the cost of production is much less."

I thanked the old schoolteacher and made my way to the door of his simple house. On one wall I noticed a glass case displaying several teaching awards, along with a framed picture of Vladimir Ilyich Lenin.

I pointed to the picture. "Are you a Communist, then?"

"Not on paper," Singh said. "But I've got a leftist ideology."

"What do you think Lenin would say about the green revolution?"

"He would say exactly what I say," Singh said. "Natural farming is the way. Like politics, you have to work from the bottom up, at the grassroots level. Otherwise, everything is poisoned."

THE HARSHEST CRITIC of the green revolution in the Punjab wasn't trained as a farmer at all. Vandana Shiva earned her master's degree in philosophy and PhD in theoretical physics before becoming India's most famous activist since Gandhi. Her books, lectures, and research efforts since the 1970s have excoriated the industrialization of food production around the globe. She holds particular ire toward the World Bank and World Trade Organization, whose free-trade agricultural policies she believes have devastated poor farmers, most of whom are women. And she has waged a bitter public battle against large biotech seed companies like Monsanto and Syngenta, which she believes are pushing expensive genetically modified seeds into India and elsewhere without regard for their safety or their impacts on local biodiversity. *Time* magazine named her an environmental "hero" in 2003, and she has won numerous awards for her efforts over the years.

I met Shiva in the cramped offices of her nonprofit Navdanya, the largest cooperative of organic farmers and seed savers in India, in a modest neighborhood on the outskirts of New Delhi. A small woman in her late 50s, Dr. Shiva entered the room in a rush of orange sari, wearing a large dark-red bindi on her forehead—the sign of a married woman among Hindus. She'd grown up on a small farm in the hill station of Dehradun, in the foothills of the Himalaya just east of Punjab. Her father was a forester, and while he was off in the woods her mother ran the farm. She attended Punjab Agricultural University and vividly remembered a happier Punjab of the late 1960s and early 1970s.

In her first book, *The Violence of the Green Revolution*, Shiva argued that much of the religious and political turmoil that had racked the Punjab from the mid-1970s onward had its roots in the growing economic disparities and unemployment that arose from the new agricultural technology. During this period a powerful Sikh separatist movement emerged that demanded an independent Punjabi-speaking Sikh state of *Khalistan*—which translates rather ominously as "land of the pure." The movement started peacefully but soon turned violent, with bombings, assassinations of government officials and journalists, and the downing of an Air India jetliner. Hindus were killed on public buses and trains. Punjabi police forces and Indian security forces responded in kind, and by the time the movement was crushed, some 30,000 Punjabis had died.

"That's six times the deaths during 9/11," Shiva said. "In Punjab every day there were killings from the '70s to 1984. The violence of the times was external. Lots of police actions. Now the violence is internalized with these suicides. In 1973 the subsidies were still there. When they started withdrawing subsidies, the troubles started."

Shiva argues that if you look at the green-revolution monocultures on a systems level, they are not as productive as traditional agriculture. All the energy in the system is focused toward producing more seed. But the short-stalked varieties leave less straw for livestock to eat, and fewer nutrients to recycle into the soil. And with less diversity on the farm, she says, there is nothing to eat should the crop fail.

"I call it 'monocultures of the mind,'" she said. "There were 250 kinds of crops in Punjab before the green revolution. It was very diverse and very mixed. Corn bread with mustard greens is the cultural identity of the Punjab."

Shiva's favorite example is a traditional green called bathua (*Chenopodium album*), a member of the spinach and chard family incredibly rich in vitamin A, calcium, protein, and potassium that has been grown and eaten in the Punjab for centuries. But beauty, as always, is in the eye of the beholder. Also known as lamb's-quarter and goosefoot, bathua is one of the most widely occurring weed species in the world. It easily outcompetes maize, wheat, and rice for nutrients, and its tall, tough stalk and prolific tiny black seeds foul up machine harvesters. Researchers have documented crop losses of up to 50 percent in US cornfields due to a combination of lamb's-quarter and giant foxtail. Bathua also loves nitrogen fertilizer. When Indian farmers began pouring on chemical fertilizers in the 1970s, bathua threatened to overwhelm the new high-yielding grain varieties. So, broadleaf herbicides were introduced to control it.

Vitamin A deficiency has long been a major health problem among Indian children and nursing mothers. In 2004, researchers at Johns Hopkins University estimated that 56 million Indian children, aged 5–15, suffer from vitamin A deficiency, including nearly 7 million who suffer some form of visual impairment—from night blindness to true blindness. Instead of harvesting a nutritious traditional food crop that could supplement the diets of the poor, Shiva says, green-revolution farmers poisoned it just to grow more grain.

EVEN SOME OF the most hardened green revolutionaries in the Punjab now recognize its costs. Dr. G. S. Kalkat, now in his mid-80s, was deputy director of the Punjab Department of Agriculture in 1964, and was one of Borlaug's staunchest disciples. Kalkat worked hand in hand with Borlaug and Dr. M. S. Swaminathan in Delhi. In the 1970s he became the agriculture commissioner for all of India, and from there he joined the World Bank as an agricultural adviser. He currently serves as chairman of the Punjab State Farmers Commission, where he has been working nonstop since 2005 to encourage Punjab farmers to diversify their crops and grow less rice.

Kalkat, a tall distinguished Sikh with a flowing gray beard, invited me into his home in a quiet neighborhood of Chandigarh, and served tea and cookies. Long ago, in 1958, he'd earned a doctorate in entomology at Ohio State. Now, in a deep baritone, he described the serious problems facing Punjab's farmers today.

"Sometime in the mid-1990s, the yield of wheat and rice on one hectare came to about 10 tons," said Kalkat, "5.5 tons of paddy, 4.5 tons of wheat. That's when the plateau came in. The inherent potential yield for these crops is 14 tons per hectare per year. Yet the current average is 10 tons for both. It's very difficult to go beyond that. Ten tons of grain, year after year, is mining large amounts of nutrients from the soil. Without large quantities of fertilizer you would not be able to produce this yield."

Shrinking farm size, Kalkat said, exacerbates the problem. In the 1970s most farmers had about 30 acres of land, and it provided them a certain standard of living. If a farmer has three sons, under the inheritance laws of India he must divide the land among them (not, however, among his daughters). So even if profit per acre has gone up, income per family has gone down as landholdings have shrunk. If the government had invested in rural industries as well as the green revolution, Kalkat believes there would be more jobs for sons off the farm.

"The government is planning to spend billions on the IT industry over the next decade, but not one rupee is going to the rural areas. We must withdraw 40 percent of farmers from the land so the remainder will have viable landholdings, a minimum of 15 acres." Some 60 percent of Punjab farmers own 10 acres or less, he said; 45 percent cultivate less than 5.

It's rare to find someone who has spent a long and successful career in an industry willing to discuss its impacts candidly. Kalkat had spent a lifetime promoting the green revolution both at home and around the globe. Yet after articulating the problems of water shortage and crop diversity, the distinguished entomologist looked at me and said, "The main problem in Punjab is the use of pesticides."

I was stunned to hear an agricultural entomologist complain about insect poisons. Typically, they are the first to rush to the chemicals' defense. If one of the architects of the green revolution had turned against pesticides, then that was truly revolutionary. "Do you think Punjab farmers should go organic then, as Dr. Vandana Shiva suggests?"

"I'm currently writing a paper on organic farming, on how far it can go," Kalkat said. "I would go organic for two things. Vegetables—we can do that—as well as organic meat, chickens, and eggs with no hormones, antibiotics, or pesticides. That we must do. But it will be very difficult to do for grains," which require fertilizer to sustain the high yields necessary to feed India's 1.3 billion people.

"Ultimately, it's a population problem," Kalkat continued, echoing Borlaug himself. "If India were a country of 700 million instead of 1.3 billion we could do fallow-land farming. Dr. Shiva has taken one line and stuck to it. I have strong disagreements with her. India is adding the population of Australia every year. Where will we get the grain to feed them all?"

India's giant neighbor to the north is wrestling with a similar, though slightly different, problem. For the last decade, China has been scouring the globe buying grain not for people, but to feed its exploding population of pigs.

Chinese workers in Guangdong Province celebrate the good times at a 13-course banquet served to 3,000 people in 2008.

CHAPTER 5

China

Landraces and Lamborghinis

> When I make enough money I'm going back to my village in
> Henan and start a pig farm. A lot of people have done this.
> As long as you have 100 pigs or more you can make a good
> living. It's a growing business.—*Sun Haipeng, 26-year-old taxi
> driver, Guangzhou, 2008*

I n the midst of the food crisis I found myself sitting in a hot-
pot restaurant in "Flower Village," a strip mall in Guangzhou,
China, growing hungrier with each passing plate. The megalopolis formerly known to the West as Canton is famous for cut
flowers, my host informed me, and the restaurant he'd chosen,
famous for entrails, specifically chicken cooked inside a pig's
stomach. Though I love Chinese food, an offal man I am not.
So I politely picked my way through five or six greasy courses
until a plate of small fried circles that looked exactly like cal-

amari was passed around the table. Ah, finally something I could eat.

I popped one in my mouth and began to chew. And chew. And chew. The taste wasn't bad, but it was a bit like gnawing on a rubber doughnut. I asked my host, an affable young importer of scientific testing equipment for the food industry named Stephen, what it was.

"That last part of pig intestine," he said flatly. "How you say in English . . . asshole?"

The joke among Chinese who live in other parts of China is that the Cantonese will eat anything that flies except a plane, anything that crawls except a tank, and anything that swims except a boat—and every part of the pig save its hair. The reason given is that for thousands of years the small, crowded province on the southeast coast of China was desperately poor and hungry, a place of frequent famines where the emperor sent disgraced officials as punishment. Mountains rim the land, squeezing its crowded cities between steep slopes and the muddy shore. With limited arable land, the Cantonese pulled sustenance from the braided Pearl River delta and the South China Sea. When that wasn't enough, they moved down the food chain to fat sandworms, big water roaches, and even a mushroom-like growth they found on rotted logs with the unappetizing name of black fungus.

Their diets and their fortunes changed dramatically after Deng Xiaoping designated the delta region as China's first Special Economic Zone in 1980, reopening Canton's legendary ports to Western trade. The province exploded with foreign investment. With a population of nearly 13 million, Guangzhou is now the third largest city in China, and migrant workers from China's hinterlands have swelled the population of the greater metropolitan area to more than 40 million. Quaint two-story houses and storefronts have given way to an endless vista of office towers, apartment buildings, and factories belching smoke into the hot, humid South China sky. But they haven't forgotten their age-old priorities. As the tour guides say, in Beijing one talks, in Shanghai one shops, and in Guangzhou one eats.

"For Chinese people, food is as important as the sky," says Stephen, who was born in the city and returned after graduating from university in Beijing. "This is especially true in Canton. We eat five meals a day—breakfast, lunch, dinner, afternoon tea, and late-night meal. And we eat anything. Snake, dog, cat, civet cat, silkworm, cricket, scorpion, rat."

"Is rat good?" I asked.

"It's better than frog."

But it's pig, not frog or rat, that is the most beloved food in all of China, accounting for nearly 65 percent of all meat consumed. The Chinese were among the first to domesticate the wild boar, some 10,000 years ago. For most of China's history, nearly every rural family kept a pig under the house as sort of a living garbage disposal for food scraps and waste—a symbol so powerful that the Mandarin character for "home" actually depicts a pig under a roof. Legend also holds that the Chinese were the first to cook the savory meat over fire: A Chinese peasant returned home one day to find his bamboo house burned down and his pig—his family's only source of boiled meat—smoldering in the ashes. When the distraught man grabbed a leg to remove the animal, he discovered the tantalizing aroma of roast pork and changed Chinese cuisine forever. In Chinese astrology, those born in the year of the pig are considered lucky, because it is said that they will always have plenty to eat.

Not surprisingly, the price of pork in the marketplace is a key indicator of Chinese zeitgeist, much like the price of gasoline in the United States. When pork is cheap, the hundreds of millions of Chinese workers who spend 40–60 percent of their income on food are relatively happy. When pork prices rise, the collective growl of a billion stomachs can rattle Beijing. Many people remember the young man with shopping bags standing down a column of tanks in Tiananmen Square in 1989. Few outside China, however, remember that the protests—which swept across many major cities and unleashed the fiercest political unrest in recent Chinese history—were preceded by a year of high food inflation and pork rationing in much of the country. The government now keeps 200,000 tons of frozen pork in a strategic reserve to help ease supplies when prices rise, just as the United States has a strategic petroleum reserve.

After Tiananmen Square, China's government encouraged pork production—through subsidies, tax breaks, and even direct payments to farmers—to stabilize pork prices and availability. In 1980 China was home to 200 million hogs. In 2005 that number had more than tripled, to 650 million. Thanks to rising incomes in China's burgeoning new middle class, per capita pork consumption has doubled since 1990, from about 20 kilograms a year to more than 40 in 2014. Chinese city dwellers eat twice as much pork as their rural cousins, and in the booming metropolises, consumption of chicken, beef, and seafood has risen dramatically as well. Though the average Chinese citizen still eats less than half the meat of someone in England or the United States, the People's Republic consumes half of the world's pork each year.

You can get a sense of China's growing pork addiction in the bustling Construction New Village Market—a dark, almost subterranean cavern jammed with stalls beneath a mountain of concrete apartment towers in Guangzhou. The orderly rows of vegetables, fish, and sacks of rice swirl with shoppers, mostly women, picking up something for dinner on their way home from work. In one row, stall after stall features glistening red meat and every cut of pork imaginable, including intestines, snouts, and ears. Butchers whack away with cleavers, carving out ribs, shoulders, and tenderloins, as blood—which you can buy by the cupful—covers the floor.

Times are good, said He Xingdong, a smiling 35-year-old butcher who had sold fresh pork in Guangzhou for a decade. "Now average people eat pork, and the rich eat seafood," he laughed. "Ten years ago I used a motorcycle to bring in my pork. Now I use a truck."

In the next stall over, 67-year-old Mrs. Quian haggled with a butcher over a choice cut. They finally agreed on a price, and while the butcher wrapped her purchase, she told me she cooks for her family of four and prepares a meat dish at every meal. She claimed she was actually trying to cut down on pork, though the slab of shoulder in her shopping bag belied the fact. "The first five years here we spent a lot of money on pork," she admitted. "But last five years we spend less, because we are worried about our health."

Though most of China still buys the freshest pork from a traditional wet market like this one, a growing number of urbanites are following their Western counterparts and buying precooked or prepackaged meals to pop in the oven or microwave at home. Even in the more traditional Construction New Village Market, a young woman manning the "Day and Day Hand-Pulled Pork" stall near the entrance was doing a brisk business in takeout, selling slow-roasted, barbecued delicacies. The mouthwatering smell wafting out of the stall took me back to the pig pickings of my North Carolina youth. "Six years ago we started with one stall," she said. "Now we have 15 all over Guangzhou."

Much of the rising demand for China's favorite white meat has been met through the efforts of animal scientists like Dr. Li Jianhao, one of China's top pig breeders at the Guangdong Academy of Agricultural Sciences. Since the late 1990s, Li has been crossing the native black Chinese pigs with faster-growing breeds from the West, including red Durocs from the United States, England's legendary pink Yorkshires, and the famous white Danish Landrace—

the modern superpigs that provide much of the foundation stock for commercial swine herds around the world.

LI PROUDLY SHOWED off the modern breeding facility tucked in the hills not far from the academy offices and labs, where 1,100 sows and 40 boars produce about 7,000 piglets a year to sell to local farmers. But he couldn't show us the breeding pigs themselves; as at most such facilities around the world, the pigs are quarantined from visitors. He did show us the big sows on video monitors, and we were able to get a glimpse of their pink-and-black-blotched off-spring snuffling about in outside pens awaiting sale.

Afterward, Li took us to a midsized commercial hog farm nearby. Great Wall Farm was a collection of about half a dozen open brick hog sheds with corrugated tin roofs tucked into a lush green valley surrounding a long lakelike hog lagoon, thick with aquatic plants. The sheds were open to the breeze, keeping the pigs cool in the tropical heat. Even with the big lagoon nearby, there was surprisingly little smell—unlike the stomach-churning stench of the hog lagoons in North Carolina. The farm had been there for years, Li explained, but only over the last decade had it expanded from less than 50 sows to more than 200.

This farm was typical of the fast-growing industry, Li explained, as small mom-and-pop operations gave way to more intensive, industrial models that employed nonfamily labor. The farm boss, an elderly Chinese man with white beard and sandals who still wore the traditional cone-shaped *cao mao* straw hat, concurred. He was hard at work at the feed mill, but he stopped to tell us that the farm had been much smaller years earlier, with no large lagoons for the waste. His son had become a successful business-man in town and had invested in the expansion. Overall, how-ever, it was an impressive place. The pens were neat and clean, and workers were busy washing things down. The pigs had their tails and ears, which are often docked on US farms, and they stood on a slatted floor where their waste could be sluiced down into catchments below that funneled it toward the lagoon. But they walked and ran about on the floors without difficulty, and the pens were not crowded.

Looks can be deceiving, however. There are two ways to increase pork production in China, said Li, and both have limitations. The first is to increase the efficiency of the small family farms that raise 50 animals or less; these still produce 70 percent of China's pork. Aside from their inefficiency—fewer piglets survive to adulthood, and those that do grow much more slowly—their open sheds and relatively close proximity to people and other hog operations makes them vulnerable to disease.

In 2005, China's swine industry was hit by an outbreak of porcine respiratory and reproductive syndrome (PRRS), also known as "blue ear disease" for its trademark symptom. As with AIDS in humans, a single-stranded RNA virus brings on PRRS in swine, causing the animals to lose their appetites, suffer fever, and develop respiratory ailments, and some eventually die. Infected sows often abort their litters or give birth to stillborn, mummified piglets. There is no effective treatment, other than culling and prevention, though Chinese scientists are desperately trying to find a vaccine. Over the next few years the disease spread through at least 25 of China's 35 provinces, devastating herds and fueling an 85 percent jump in pork prices in 2007.

Agricultural officials downplayed the impact, but a 2011 report by the US International Trade Commission estimated that China had destroyed 10 million infected hogs between 2007 and 2008. Even without such pandemics, Chinese farmers lose an estimated 25 million hogs to disease every year, and the number of piglets in each litter has declined as well. China's sows are now about half as productive as those in the United States.

The other way to increase productivity, Li said, is to follow in the footsteps of the United States and Europe and build more CAFOs, the controversial "concentrated animal feeding operations" that raise pigs from birth to slaughter weight in tightly confined pens indoors. Though highly efficient at turning soybeans into bacon, such factory farms have unleashed a firestorm of protest from animal welfare groups in the United States and Europe over living conditions that would be illegal for cats or dogs, and over the routine use of antibiotics that has given rise to resistant strains of bacteria.

Few such concerns have been raised in China. After Tiananmen Square, the Chinese government gave tax breaks to companies willing to set up CAFOs, and nearly a thousand have been built that each produce more than 10,000 hogs annually. But even Chinese swine experts doubt that the country can build many more, because of the lack of available land and water. Each year, China tries to feed four times the population of the United States on roughly a quarter of the arable land. Most of China's best agricultural lands are in the east—the same area that has seen rapid urbanization. Given the waste and water problems associated with CAFOs, local officials much prefer factories—which produce more jobs and revenue—over farms. One Chinese agricultural expert estimated that a 10,000-pig CAFO would require at least 200 Chinese acres (about 33 US acres). Since most Chinese peasants farm just half of a Chinese acre, cobbling together enough land would require negotiating with hundreds of farmers.

Guangzhou Lizhi Farms, built in 1997 as a joint venture between investors in Hong Kong and Macau, is the largest CAFO in Guangdong, with 20,000 breeder sows that produce 80,000 hogs per year. The taxi driver found it with difficulty at the end of one of Guangdong's brand-new four-lane highways. Behind a large gate off a substantial gravel road stood 60 hog houses and several waste lagoons covered with aquatic plants. The 60-hectare facility boasted a water treatment plant large enough to handle the sewage of a small town.

The farm's technical manager, Qingzhang Lu, spoke to us in the conference room of the CAFO's large office building beneath a glaring picture of Chairman Mao. He was a sharp young man who wore round John Lennon glasses and showed us a PowerPoint presentation describing the farm. He said they hadn't been hit by blue ear disease yet, but he wouldn't let us anywhere near the hog houses without a disinfecting shower and 48 hours of quarantine. He was confident that despite the constraints of land, water, and rising grain prices, China could meet its skyrocketing demand for pork. "There's still room for Chinese pig development," he said. "We just need to set up more CAFOs."

Even with more CAFOs, the farmers will still need to feed the pigs, and therein lies one of the greatest challenges for China and the rest of the world. In 1995 a seismic shift in the global grain trade began as China went from a soybean exporter to a soybean importer, primarily to feed its half-billion hogs and provide cooking oil to its 1.3 billion people. China started importing US corn and soybeans and then began buying soybeans from Brazil and Argentina as the soy revolution rolled into South America. Within a few years China became the world's largest importer of soybeans. Soy imports have grown, from about 10 million metric tons in 2000 to nearly 60 million in 2012. China's domestic soybean production of about 14 million tons has been flat or falling as the government has encouraged farmers to plant more profitable corn to feed cows and chickens. Yet annual consumption is 70–72 million tons and rising. The nation now consumes more than 60 percent of global soy exports—nearly twice as much as all other soy importers combined. Another seismic shift in China's grain supplies occurred in 2007, when the nation went from a longtime exporter of corn, rice, and other grains to a major importer, pur-chasing 18 millon tons of grain on the world market in 2013. That year, 70 percent of China's wheat came from the United States. The trend shows no sign of slowing.

Shen Guang Rong has ridden the pig boom since its inception. He came from a tiny village near the Russian border in 1982 to work at China's first CAFO in Shenzen when the town was little more than a fishing village. It's now the third-largest container port in the country, with 10 million residents, famous for produc-ing not pigs, but iPhones, Segways, and skateboard decks, among thousands of other consumer products. Shen's former employer—one of the largest meatpacking companies in China—was the first to import US corn for hog feed in the early 1990s. In 30 years Shen has watched the number of CAFOs skyrocket. Today he is a top consultant to the CAFO industry, with a tenth-floor office overlooking the megalopolis.

"In 2008, as long as you kept a pig you made money," Shen said from behind his sprawling desk, where two giant computer screens flickered the latest commodity prices. "Many foreign investment banks invested in Chinese pork production. Goldman Sachs, Deutsche Bank, the big Thai bank Krungthai. Even local stock-brokers and developers were investing in pig farms."

But Shen doesn't think that sort of investment will last, because the profitability of pigs hasn't changed much over the years, he says. It's one of the few industries that nearly 700 million rural Chinese peasants can jump into or out of easily when prices go high or low. Shen also downplayed the consumptive demand from China's growing middle class. Instead he pointed out that the fac-tories must feed a certain quota of meat to their migrant workers, which has driven consumption in cities. "Everybody talks about China as a market of 1.3 billion," Shen said. "But that's crazy, since 80 percent of Chinese do not have the consumption power."

The consumptive power of China's factory owners, however, was evident in Guangzhou. After visiting the lovely traditional pagoda-like Sun Yat-sen Memorial Hall honoring the father of modern China, I walked around a street corner and did a double take. A neon-orange Gallardo Spyder glowed behind the glass of a Lamborghini dealership, beckoning Guangzhou's nouveau riche with its polished curves, air dams, and V-10, 520-horsepower engine. I entered the showroom to get a better look, and a svelte saleswoman clad entirely in black assured me that the orange Spyder could catapult the modern capitalist comrade from 0 to 60 miles per hour in a blistering 4.3 seconds. She claimed they sold several of the nearly million-dollar machines every month.

An audacious, screamingly fast Italian sports car seemed a strange but somehow appropriate symbol of China's new wealth, which is growing so fast it's always on the verge of spinning out of control. But it suited Shen Guang Rong just fine. Wearing loafers, a gold Rolex, and a pink polo shirt emblazoned with the Playboy bunny, Shen looked more like a Las Vegas high roller than a man who made his money in swine. He graciously invited me, my translator, and the translator's friend Stephen, the importer of scientific equipment who had driven us to Shenzen that day, to lunch at his favorite restaurant in the building next door, a shimmering blue glass office tower with so many flags flapping out front that it looked like UN headquarters. Waitresses clad in gold silk dresses seated us in a private room and proceeded to bring out course after course of steaming dishes. Glasses of Chinese beer and rice wine were drained around the table and quickly replenished. After two hours or so, the conversation between the Chinese men was flowing as fast as the wine.

"Aren't you concerned that China may not be able to meet the rising demand for pork because of your dependence on foreign grain imports?" I interjected. Shen smiled and then spoke slowly, as if addressing a dim-witted child. "China is a very special country," he said. "Throughout Chinese history China had a tradition of self-sufficiency. Even though now China relies on many imports of grain and raw materials, China will not rein in prosperity as long as we are peaceful. But China always has a bottom line. If there is a war with the US, it will not be a problem to feed its people. It just depends on what we feed our people. Now Chinese drink lakes of alcohol. We could stop drinking and eating meat and we'd still be able to feed ourselves."

It sounded a bit like the chain smoker who claims he can quit any time. Yet the idea of war with China—and the ease with which Shen had brought it up—sent a shiver down my spine. In retrospect, he was probably right. It would probably take war, a worldwide economic depression, or at least a prolonged drought for China to voluntarily resume its low-meat diet of 1979.

The Chinese government is much less sanguine. The lure of jobs and better pay in the cities has become such a powerful draw for rural migrants that Beijing has made rebuilding the countryside a top priority in the last two Five Year Plans, raising farm subsidies to boost stagnant food production.

On the way back to Guangzhou, we took a detour to see the port of Xinsha on the Pearl River, where bulk cargo ships the size of aircraft carriers were disgorging their bowels into giant grain elevators that lined the bank. Many belonged to huge soy-processing facilities—known as "crush plants"—that make Guangdong the livestock feed capital of China. We drove past the State Grain Reserve, a complex of green-and-white silos rising in stately rows connected by a maze of elevators and conveyor belts that spanned a city block. Right next door was an equally expansive facility with a familiar name: Cargill.

Based in Minnetonka, Minnesota, the agricultural behemoth has become one of the largest suppliers of grain, meat, processed food, and biofuels to the world. The sprawling Cargill complex had "sprouted from the Machong banana fields" in 2003 and could crush 3,000 metric tons of soybeans a day. In 2005 Cargill bought the crush plant next door, adding another 1,800 tons of capacity. Cargill does more than half of its business in developing countries, where it vigorously promotes the green-revolution equation: free trade plus industrial agriculture equals rising incomes for people that can afford more meat, as well as packaged and processed foods. All of these Cargill happily supplies—from the ships that bring in soybeans from Brazil and the United States, to livestock feeds, to frozen pork, beef, and chicken, to fast-food burgers, to chocolate. Rising food prices have benefited Cargill and its kind. In 2011—as most other industries were still struggling to recover from the recession—privately held Cargill posted a profit of almost $3 billion, a 35 percent increase over the previous year.

Port workers, some in dust masks, hustled to catch shuttle buses home in the midst of a thick brown haze—a potent cocktail of ship exhaust, dust from the giant augers and conveyors, and smoke from the crush plants that carried a fragrant mist of soy oil through the air. The sun was a dirty orange ball sinking below the horizon. It had been a long day, and our hotel in Guangzhou was still hours away. And in typical Chinese fashion, the road we needed to get back out to the highway was under construction. We crept along in bumper-to-bumper traffic on a slightly elevated two-lane blacktop interspersed with long stretches of washed-out, potholed roadbed that detoured us into the adjacent fields. One of the big trucks hauling a shipping container inevitably got stuck on the dirt-track detour and brought the entire lumbering parade to a dusty halt. We waited for over an hour. Finally, since Stephen owned an SUV, we decided to head off on a parallel dirt track that spun off to a workers' village in the distance. Row after row of high-rise concrete dorms clustered around a small commercial center where workers could buy cell phones and counterfeit brand-name clothes. We parked next to a sidewalk café to grab a bite and wait for Chinese officialdom to pull the truck from the mud.

Three young men in clean, pressed shirts sat at the table next to us drinking beer and relaxing after work. We introduced ourselves and learned that they were all mechanics at the gargantuan Nine Dragons paper mill that we'd passed in Dongguan on the way to the port—one of the largest paper mills in the world. The plant and its CEO, Zhang Yin, China's richest woman, had recently been criticized by labor groups for numerous labor violations. The "paper empress" had also been pilloried in the Chinese press for zealously defending China's yawning income gap and advocating lower taxes on the rich and fewer legal protections for workers.

The men were all migrants from various rural regions of China. Mr. He, 25, hailed from Guangxi; Mr. Liu, 23, came from Hunan; and Mr. Deng, the only married man in the group and the elder statesman at 37, came from Sichuan. I asked them what had brought them to Guangdong Province.

"My mouth brought me here," Mr. Liu said with a laugh. Here he made more than 2,000 yuan a month, double a mechanic's wage back home. "And I get to feed myself," he added.

Mr. Deng, who had been in the city the longest, was more phil-osophical. "City life has advantages and disadvantages," Deng said. "The environment is better back at home, while in the city there are a lot more things to see and you make more money. Some things are cheaper here, like electronics. But the vegetables are not as fresh. There is not as much variety in the food as back home. But of course we eat more meat here. I've actually gained a little weight."

Each man wanted to make enough money in the city to go back home to set up a business. Mr. He wore a bright-green shirt, sported mod rectangular glasses, and had a sharp, intelligent gaze.

He seemed like a young man on a mission. As we got up to leave, he pointed to Stephen's vehicle, a silver Mitsubishi SUV. "I want to have a car like that someday," he said.

It was a simple statement. Something that would be unremarkable made anywhere in the West. I wanted a car like that when I was his age too, and I worked hard and eventually bought one. Yet here it seemed to encapsulate the current world conundrum. I have no doubt that someday He, and millions of other hardworking Chinese like him, will have the cars they want. But what those collective, middle-class aspirations portend for the world's food, water, energy, and climate is the 9-billion-person question.

Providing pork for China's factory workers and growing middle class—as well as the grain to feed all the pigs—is the more immediate challenge. Like India, China has made huge strides in grain production using the tools of the green revolution. Yet like India, grain production since 2000 has plateaued; in some years it has fallen abruptly, largely because of drought. In 2010, China imported 95 million metric tons of grain—17 percent of its domestic production, and more than the average annual grain harvest of Russia.

China's grain addiction has had a ripple effect around the globe. It's helped to drive a rapid expansion of industrial-scale soybean production in Brazil, which has fueled the deforestation of the Amazon and the plowing up of the adjoining Cerrado, Brazil's great grassland ecosystem. Between 2003 and 2008, Brazil's soybean acreage increased by 39,000 square kilometers, an area larger than the Netherlands. Since 2000 the rate of deforestation in the Amazon has risen in lockstep with the price of soybeans. Major multinational grain conglomerates like Cargill, ADM, and Bunge have facilitated the great plow-up, building new roads, river ports, and fertilizer distribution facilities to help boost soy exports from the region.

"I don't remember the year that China's population will reach 1.5 billion," Dr. Chen Yaosheng, one of China's top swine scholars, told me during an interview at his lab at Sun Yat-sen University. "But based on the current numbers, by then annual per capita pork consumption will reach half a pig (45 kilograms). Which means China will need 750 million pigs. If you take urbanization into consideration, China will have to produce 800 million head a year to meet that demand." Feeding that many more pigs alone will be a formidable challenge. But thanks to their newfound wealth, the Chinese are also eating more beef, poultry, and farm-raised seafood, all of which are fed with grain, among other ingredients. In 2009 the Chinese government renewed its policy to be 95 percent self-sufficient in cereals and released a national plan to grow 50 million more tons of grain by 2020.

Instead they are headed in the opposite direction. Dr. Chen's lab was in the spanking new Guangzhou University City, an amalgamation of 10 universities built on 10,000 acres of former farmland on an island in the middle of the Pearl River. The megacampus—built in just five years to accommodate a quarter-million students, faculty, and staff—represents China's future: book-toting, pop-culture-loving future scientists, factory owners, and engineers. Yet as we left the island I caught a glimpse of another China: in a small field almost overshadowed by the elevated interstate, a stooped woman wearing a *cao mao* was slicing at the red-clay rows with her wooden hoe. I asked our taxi driver to pull up next to her plot. Even today, despite its efforts to modernize and industrialize its food production system, China is not so much farmed as intensively gardened by people like Ms. Li Xiufen. She was 50 years old, a widow, and had been farming this Chinese half acre (1 mu = 0.16 acre) most of her life, she said. All her children had gone to the city. She usually planted sweet potatoes and rice, but this year the village leader had asked her to plant bananas. She used to keep pigs as well, she said, but the village didn't allow it anymore because of the smell and waste problems.

Ms. Li and her comrades-in-hoes fed China for thousands of years. But the disparity in income, diet, and quality of life between urban and rural China is so great now that the Chinese government is having trouble keeping farmers on the land. Even many of those who want to stay have been kicked off by corrupt local officials, who can get rich quick on kickbacks by leasing the land to factories or developers. The Chinese government estimates that 40 million farmers have been displaced this way, though other estimates place the number at 80 million, with another 2 million displaced each year.

The government has long maintained that the country must keep at least 120 million hectares of farmland to ensure food self-sufficiency, but the Chinese are caught in a farmland crunch. Development, erosion, and desertification take a million hectares of farmland out of production each year, and the quality of much of the remaining soil is questionable. In 2007, former land minister Sun Wensheng warned that at least 10 percent of China's arable land was contaminated with heavy metals and other toxic pollutants. More than 3 million hectares of farmland has been deemed too toxic to grow crops on. In early 2011, *China Economic Weekly*, a magazine controlled by the Communist Party, reported that 12 million tons of grain had to be destroyed because it was contaminated with cadmium, a known carcinogen, as well as other metals. As much as 60 percent of the rice sampled in several southern provinces exceeded China's national standard for the toxic metal. Today, half of China's rivers, a major source of irrigation water, are so polluted that their waters are unfit for human contact.

AS A LAST resort, China has increasingly turned to the United States and other countries to meet its growing pork demand.

During the 2008 Olympic Games in Beijing, blue ear disease was still ravaging Chinese swine herds, so officials imported 372,000 tons of frozen and chilled pork directly from the United States. It's not the favorite of Chinese consumers, who still prefer local varieties as fresh as they can get them. But it helped ease some of the supply and price concerns. In July 2011, when China was hit by a 57 percent jump in pork prices, again the government turned to imports of frozen pork from the United States, which by the third quarter of 2011 had hit a record 445,000 tons, along with more than 620,000 tons of offal, offcuts, and organs. In a sign of China's increasing dependence on frozen-pork imports, in 2013 China's biggest pork producer, Shuanghui International, bought Smithfield Foods, the largest hog producer and processor in the world, for $4.7 billion. The deal was the largest Chinese takeover of a US company in history. Some US consumer groups raised concerns about eating tainted pork from China. The far more likely scenario is that more and more US pork will be heading in the other direction.

Even if China imported all the world's available pork, however, it would make only a slight dent in the country's demand. The total world pork export market—dominated by the United States, Canada, and Denmark—is about 6 million metric tons. Each year, China consumes some 50 million metric tons of pork—enough to fill nearly 2 million standard-sized shipping containers. Placed end to end, they would create a flotilla of pork stretching across the ocean from San Francisco to Shenzen and back.

The real potential of a major pork and grain shortfall terrifies analysts both in China and overseas. David Daokui Li, a prominent economist and member of the powerful Monetary Policy Committee of the People's Bank of China, warned in 2011 that China was headed toward the "perfect storm" of grain. A recently declassified report by the US National Intelligence Council (NIC), echoed Li's concerns. According to the NIC report, China has experienced more frequent droughts, floods, and major storms in recent years than in all its long-recorded history, causing economic losses of $25–$37 billion each year. The spy agency concluded that a water crisis in China was all but inevitable and could have major repercussions in the country.

Li's recommendation to China's leaders was to invest heavily in agricultural technology to boost yields at home—as well as in agricultural lands overseas. They are heeding his advice. In the past decade, China and Chinese companies have been quietly gobbling up millions of hectares of farmland in Africa, South America, Ukraine, Australia, the Philippines, and elsewhere as an agricultural lifeboat for the coming storm. Other wealthy, land-poor countries like South Korea and Saudi Arabia are doing the same.

ON MY LAST night in Guangzhou, my translator took me to a famous seafood restaurant on the banks of the Pearl River. As we strolled along the promenade, I couldn't help but reflect on the river's role as an epicenter of world trade for nearly four centuries. Spanish galleons sailed these waters with holds full of South American silver that they traded for silks. Britain's squat East Indiamen and sleek American clippers anchored off its mouth in darkness, their sailors unloading countless chests of illegal opium to the junks of Canton smugglers. From almost the beginning of the mercantile era, the West could not do without the riches of the East.

The water's inky blackness now reflected the glaring neon bling of the new city, rising like Las Vegas from the marshy delta to flash its new wealth to the world. The bridges blazed with multicolored lights. Buildings sported Times Square–ish jumbotrons flickering ads over the city. Large tour boats cruised up and down the channel lit up like a Christmas flotilla. One even carried a two-story electronic billboard that flashed commercials for luxury goods to those walking along the bank—Rolex, Gucci, Prada. It seemed every retailer in the world was after China's newfound urban consumer as the country did its best to out-East the West. Guangzhou at the moment, like China itself, is a booming, sprawling money machine running on domestic coal and imported soybeans. Just as the emperor couldn't stop the opium trade, China's current leaders are powerless to break their nation's growing pork addiction.

When we reached the restaurant, my translator poked me in the arm and pointed. Parked in the VIP space just a few meters from the door stood a smoke-gray Lamborghini.

ONE PROMINENT CHINA watcher, however, foresaw the nation's food troubles long before they reached the current precarious stage. "An immense capital could not be employed in China in preparing manufactures for foreign trade without taking off so many labourers from agriculture as to alter this state of things, and in some degree to diminish the produce of the country. The demand for manufacturing labourers would naturally raise the price of labour; but as the quantity of subsistence would not be increased, the price of provisions would keep pace with it."

His name? T. R. Malthus.

A worker takes a break amid a sea of sugarcane in São Paulo state, Brazil, once dubbed the "Saudi Arabia of ethanol."

CHAPTER 6

Food, Fuel, and Profit

Alcohol can be manufactured from corn stalks, and in fact from almost any vegetable matter capable of fermentation. . . . We need never fear the exhaustion of our present fuel supplies so long as we can produce an annual crop of alcohol to any extent desired.—*Alexander Graham Bell, 1917*

When Dario Franchitti steered his sleek, black, 650-horsepower Indy car into the winner's circle after the rain-soaked 2007 Indianapolis 500, the soft-spoken Scotsman drank the ceremonial buttermilk; kissed Ashley Judd, his movie-star wife; and was probably too consumed by his first Indy 500 win to consider the record he'd just set in racing history—or the ripple effect it would have on millions of starving people around the planet.

That day Franchitti became the first driver to win America's century-old auto race on 98 percent ethanol—the 113-octane, gin-clear grain alcohol distilled from corn grown right there in the Hoosier State. Race officials had planned to use pure ethanol until they discovered they would have to pay a steep liquor tax on thousands of gallons of drinkable fuel. So they added 2 percent gasoline to the mix.

Robert Malthus and Amartya Sen may have disagreed on the causes of famine and hunger, but for both the solution was practically the same: produce or import more food. More food in the market pushes prices down during a market failure and makes more calories available to more people during an absolute dearth. But at the turn of the twenty-first century, two new factors drastically changed that simple equation: food grains became the feedstock for a new global industry, and commodity markets were deregulated, enabling investment banks and brokerages to buy and sell billions of bushels with a keystroke. No matter how much grain is produced, poor hungry people around the world must now compete for it with Big Biofuel and the wolves of Wall Street.

Franchitti's Indy win in 2007 was perhaps the pinnacle of what would become known as the biofuel decade. It was a media stunt drummed up by ethanol producers and Indy circuit promoters in hopes of convincing a skeptical public that ethanol was a bona fide motor fuel while at the same time greenwashing auto racing's oily image. The race had been run on "wood alcohol," or methanol (made from natural gas), since the sixties. But who could blame them for jumping on the biofuel bandwagon? In 2007, everyone from Willie Nelson to farm-state politicians was singing the praises of corn ethanol and soybean biodiesel. Such homegrown fuels would bolster the rural economy, help us kick our lethal addiction to Middle Eastern oil, and reduce this country's ballooning greenhouse gas emissions. Unlike the ancient carbon released from the burning of fossil fuels, the carbon in corn ethanol and soybean biodiesel is captured by crops during the growing season. In theory, burning a tankful of ethanol could make even a 2-mile-per-gallon Indy car carbon-neutral.

The reality, however, is far less rosy. Corn ethanol contains only 66 percent of the energy of gasoline, which means that cars and trucks running on E10—the 10 percent ethanol blend now in US gas pumps—actually get about 4 miles per gallon *less* than they would if they burned straight gasoline. Nor is corn ethanol very efficient to produce. It takes almost as much fossil-fuel energy to grow corn, harvest it, haul it to the ethanol plant, ferment it, distill it, and truck it to gas stations as ethanol actually displaces. Both the fermentation process and the manufacture of nitrogen fertilizer emit copious amounts of carbon dioxide—and corn uses more nitrogen fertilizer than any other US crop. Though the ethanol industry claims that the fuel reduces direct greenhouse gas emissions by 59 percent compared to gasoline, the EPA calculates a far more modest 7–32 percent emissions reduction, depending on yields, the type of tillage or irrigation used, and the fuel burned to fire the boilers of the ethanol plant's giant stills. If the stills are coal fired, as they are in a few older plants, ethanol's greenhouse gas emissions can actually be worse than those of gasoline. For newer plants fired by natural gas, the EPA calculates about a 21 percent reduction in emissions, though even that number is optimistic.

Biodiesel fares a bit better. In the United States, most biodiesel is made by filtering and chemically treating soybean oil to get rid of the fats in a process called "transesterification." In Europe, where half the cars run on diesel, biodiesel refineries use rapeseed (the source of canola oil) or palm oil imported from the tropics, typically Malaysia or Indonesia. The CEO of a biodiesel company once told me that olive oil makes the highest-quality fuel but is far too expensive to burn in cars.

As far as efficiency goes, biodiesel contains about 90 percent of the energy of petroleum diesel and produces 2–3 units of energy for each unit of fossil fuel required to make it (compared to just 1.4 units of corn ethanol produced from a unit of fossil fuel). Biodiesel's greenhouse gas emissions are nearly 60 percent lower than those of standard diesel fuel. But again the environmental benefits depend on how and where the crop is grown. A palm-oil plantation on newly cleared peat lands in the dwindling rain forests of New Guinea produces biodiesel with a greenhouse gas footprint equivalent to that of oil boiled from the Alberta tar sands.

The United States has been called the Saudi Arabia of corn. We produce 36 percent of the world corn crop and supply nearly half of the world's exports. But even the United States doesn't grow enough corn to put much of a dent in our astronomical transportation fuel consumption, which amounts to more than a quarter of the world total. A bushel of corn yields only about 2.8 gallons of ethanol, and a bushel of soybeans yields only 1.5 gallons of biodiesel. We, the SUV-loving people of the United States burn through 134 billion gallons of gasoline and 36 billion gallons of highway diesel fuel each year. Even if you turned every kernel of the record-breaking 13-billion-bushel 2009 US corn crop into ethanol and every soybean grown in the United States (3 billion bushels in 2011) into biodiesel, they would supply only 27 percent of our annual gasoline use and a paltry 12 percent of our diesel consumption. Since corn and soy are the primary feed grains for cows, pigs, and chickens, the endeavor would also eliminate most of the beef, pork, chickens, eggs, and dairy products from our tables.

There are other problems as well. About 100 million older cars on the road today (40 percent of US cars in 2011) were built before 2001 and can't withstand the corrosive effects of more than 10 percent ethanol in their fuel systems. Using gasoline with more than 10 percent ethanol voids their warranties. The liability issues associated with older cars, along with the billions of dollars required to build ethanol pipelines from the Corn Belt (where most ethanol is produced) to the coasts (where most US cars are), as well as to install higher-percentage-ethanol pumps in gas stations, mean it will be extremely difficult for ethanol to exceed the current 10 percent of the fuel supply. As a result, the corn ethanol industry has hit what's known as the "blend wall."

So why are we now planting more acres in corn since World War II, and turning nearly 40 percent of the harvest (which would feed everyone in Africa for a year) into a motor fuel that gives less bang for the buck than petroleum fuels, has limited environmental benefits, and can never significantly reduce the amount of oil we burn? The short answer is pretty simple: money. Lots and lots of money flowing from taxpayers, grocery shoppers, and gasoline consumers into ethanol producers' and corn farmers' pockets.

The long answer involves nearly a century of federal agricultural and energy policies. Both have conspired to create an entirely new market for food grains that is ratcheting up food prices, destroying rain forests, taking up arable land, and displacing small farmers around the globe. At the height of the biofuel boom in 2007, Jean Ziegler, the former United Nations Special Rapporteur on the Right to Food, articulated the views of many groups working to alleviate global hunger: "It is a crime against humanity to convert agricultural[ly] productive soil into soil which produces food stuff that will be burned into biofuel."

HUMANS HAVE BEEN brewing grains, fruits, and vegetables into alcohol for almost as long as we've been growing grain. Archaeologists have dug up beer jugs dating to the late Stone Age, about 10,000 BC, and the Egyptians were worshipping Osiris, god of wine and inventor of beer, by 4000 BC. And we've been burning alcohol in our vehicles since the invention of the automobile. When New England inventor Samuel Morey built the first internal combustion engine in the United States in 1826, he ran it on a combination of ethyl alcohol and turpentine. In 1860, German inventor Nicholas Otto ran the first "Otto-cycle" engine—the precursor to modern gasoline engines—on ethyl alcohol. Even Henry Ford's first mass-produced car—the 1908 Model T, a.k.a. "Tin Lizzie"—was a flex-fuel vehicle, designed to run off gasoline or alcohol with controls right in the cockpit to adjust the carburetor and spark advance for optimal performance on either fuel. Early tractors in Europe and the United States ran off alcohol, often produced by stills right on the farm. When Rudolf Diesel introduced his powerful and highly efficient heat-combustion engine at the Paris World's Fair in 1900, he ran it on peanut oil.

Henry Ford, the son of a farmer, was an outspoken booster of alcohol fuels. He was a prime sponsor of the "chemurgy" movement in the 1930s to produce fuels and other useful industrial products from crops like grain and even hemp. Two major goals of the movement were to help the rural economy of the Midwest during the Great Depression and—in what has become a constant political refrain ever since—to reduce our dependence on imported oil. There was even significant research as early as 1920 into producing cellulosic ethanol from crop residues such as logging waste and even seaweed. Professor Harold Hibbert, who conducted the research at Yale University, was a man ahead of his time. "In . . . 10 to 20 years this country will be dependent entirely upon outside sources for a supply of liquid fuels," Hibbert wrote in 1920, "paying out vast sums yearly in order to obtain supplies of crude oil from Mexico, Russia and Persia."

The discovery of substantial oil fields in Texas, Oklahoma, the Gulf of Mexico, and California turned the United States into the largest oil producer in the world and made gasoline far cheaper than alcohol, ending ethanol's early fuel run. Corn fuel staged a small comeback, however, during the 1970s energy crisis. The yearlong OPEC oil embargo in 1973, followed by the 1979 Iranian Revolution, led to soaring gas prices, long lines at US service stations, and high inflation. By the end of the decade, US ethanol plants were turning just under 1 percent of the US corn crop into about 20 million gallons of fuel alcohol to blend into "gasohol," the same 10 percent ethanol mix we have today. But not enough was produced to reduce prices or the lines at the pump.

It was also during the 1970s that a small, relatively obscure midwestern businessman named Dwayne Andreas began exerting a Napoleonic influence on politicians in Washington, DC. Andreas was a poor Mennonite farm kid from Iowa who dropped out of college to trade corn, made his first million at 27, and eventually transformed a modest, family-owned Minnesota grain company into one of the largest agribusinesses in the world—now the Decatur, Illinois–based Archer Daniels Midland. ADM had three major businesses under Andreas: it sold and shipped US grain to hungry foreign nations, it processed corn into high-fructose corn syrup and other foods for consumption at home, and it produced roughly two-thirds of the ethanol in the United States from the 1970s until about 2000. During those three decades Andreas, along with his family, his company, and the company's subsidiaries, collectively became one of the largest—if not the largest—sources of political cash in the United States, doling out millions to candidates from Richard Nixon to Bob Dole to Bill Clinton.

Thanks in no small part to Andreas's political clout, Congress and President Carter passed the Energy Tax Act of 1978 that exempted gasohol containing at least 10 percent ethanol from the federal excise tax on gasoline—essentially creating a subsidy of 40–60 cents per gallon. At Andreas's urging, Congress also slapped a 54-cents-per-gallon tariff on ethanol imports in 1980. The tariff was targeted specifically at Brazil, the world's largest ethanol producer at the time, which makes its ethanol from sugarcane. A ton of sugarcane produces nearly eight times the ethanol that a ton of corn produces, and Brazilian sugar barons could make it far more cheaply.

By 1995, newspapers and magazines from across the political spectrum—the *New York Times*, the *Wall Street Journal*, *Mother Jones*—were decrying the government largesse poured onto ethanol producers, particularly Dwayne Andreas and ADM. A report by the Libertarian Cato Institute estimated that between 1980 and 1995, US taxpayers and consumers had spent nearly $10 billion subsidizing ethanol—the bulk going into ADM's coffers. That number had doubled to $20 billion by the time Congress finally ended the subsidy and the tariff at the start of 2012.

Even with cheap corn, heavy government subsidies, and protection from imports, ethanol couldn't make it as a mainstream motor fuel. But it does have a few big benefits over gasoline. It has an octane rating of about 113, equivalent to racing fuel, and burns as clean as a Bunsen burner, with little ash or smoke. It's also virtually nontoxic unless you drink too much of it, and if you spill it, it evaporates or degrades quickly, with little environmental damage. For these reasons, ethanol was used as an octane booster for gasolines during the early days of the automobile, when gasoline quality was poor. But it was soon displaced by a far cheaper additive produced by the oil industry: tetraethyl lead. From the 1920s to the 1970s, leaded gasoline became the greatest source of atmospheric lead pollution on the planet, and one of the greatest public health threats in US history.

Just as the government was phasing out leaded gasoline in the 1970s and phasing in fuel ethanol, the oil and gas industry came up with another clean-burning, octane-boosting gasoline additive, known as methyl tertiary-butyl ether, or MTBE, made from a waste product of the oil-refining process. The oil and petrochemical industries began blending MTBE in gasoline in 1979, and for the next 20 years they vigorously promoted it as a way to boost octane, help engines burn cleaner, and clear smoggy urban air.

There was just one small problem. Any improvement in air pollution from the use of MTBE was more than offset by the nationwide groundwater contamination it caused, which began just a year after it was first introduced. Oil company engineers learned that the chemical, which smells and tastes like turpentine, readily dissolves and spreads in groundwater. Once there, it breaks down very, very slowly. Shell engineers soon came up with new acronyms for the compound, including "Most Things Biodegrade Easier" and "Menace Threatening our Bountiful Environment." Even though the chemical causes cancer in rats, and EPA considers it a potential human carcinogen at high doses, from 1992 to 2002 the amount of MTBE added to the US fuel supply nearly tripled. Affected cities and communities eventually began suing the oil industry for contaminating their groundwater, giving MTBE a new acronym among oil industry personnel: "Major Threat to Better Earnings." California banned MTBE in 2002, and other states soon followed suit. The Clean Air Act, however, still required gasoline refiners to blend gas sold in smoggy areas with a clean-burning additive (technically, an oxygenate like ethanol). As more and more states banned MTBE, they ultimately created a 10-million-gallon-a-day hole in the nation's fuel supply for ethanol to fill.

The banning of MTBE ushered in the "biofuel decade." From 2000 onward, US ethanol production—and the amount of corn diverted to make it—doubled every three or four years, from 1.6 billion gallons in 2000 to more than 13 billion gallons in 2010. As usual, fuel alcohol had plenty of government help. In 2005 Congress passed a new Energy Tax Act, which, despite heavy lobbying from the oil industry, failed to shield the companies from MTBE lawsuits, effectively ending the chemical's use in the United States. (Despite its horrific environmental record, MTBE is still produced in the United States and shipped overseas, mainly to China.) The congressional act also created a federal windfall for ethanol that even Dwayne Andreas himself couldn't have imagined. This was the year that back-to-back Hurricanes Katrina and Rita sank more than a hundred oil platforms in the Gulf of Mexico, shutting down 90 percent of the region's oil and gas production—nearly a third of domestic oil production—and starting the historic climb in oil prices.

Not only did the 2005 Energy Tax Act remove ethanol's chief competitor from the marketplace, but it actually *required* oil refiners to blend increasing amounts of biofuels into US gasoline or face steep fines. Known as the Renewable Fuel Standard, or RFS, the initial mandate called for the blending of 4 billion gallons of renewable fuel into US gasoline by 2006, and 7.5 billion gallons by 2012. The law proved so popular among corn-state politicians that in 2007, Congress expanded the mandate and added a provision for nongrain biofuel. But despite billions of federal and venture-capital dollars invested in research and development since the 1970s, cellulosic fuels brewed from corn stover, wood chips, switchgrass, and the like have failed to materialize in any significant quantity. Of the 16 billion gallons of renewable fuel used in 2013, more than 80 percent came from corn ethanol.

The biofuel mandate quickly created huge demand for corn. For most of the 1990s and the early 2000s, the wholesale price of ethanol was pegged at about 50 cents per gallon higher than the cost of unleaded—the cost of gas plus the excise credit—and the two fluctuated almost in lockstep. In 2001, when corn was selling for about $1.60 per bushel and ethanol was trading at $1.77, ADM could take $1.60 worth of corn, turn it into 2.8 gallons of ethanol, and sell it for almost $5. The return on investment for ethanol producers approached 50 percent. Ethanol prices surged further as oil went from $60 per barrel in 2005 to $145 in the summer of 2008. ADM soon had a lot of competition in the ethanol game. Everyone from billionaires like Sun Microsystems cofounder Vinod Khosla and Virgin Atlantic founder Richard Branson to farmer-owned cooperatives throughout the Midwest were building ethanol plants as fast as they could. At one point in 2007 there were so many ethanol plants on the drawing board in Nebraska that the second-largest corn-producing state in the country would have had to import corn to supply them all. At the end of the decade, the United States was the largest ethanol producer in the world, with nearly double the production of Brazil, its closest competitor.

Europe was not immune to the biofuel craze; some European countries mandated biofuel use as early as 1992. In 2001 the European Union set a goal of replacing 5.75 percent of all transport fuels with biofuels by 2010, and 10 percent of its transport fuels by 2020. It also protected its producers by placing a $1.10-per-gallon tariff on imported ethanol and an ad valorem duty of 6.5 percent on imported biodiesel. Several EU countries, particularly Germany, provided generous tax breaks and mandatory blending limits for biodiesel and ethanol fuels as well. In all, more than 40 countries enacted biofuel mandates or targets, following the lead of the United States and the European Union, which together in 2006 spent $10.5 billion on biofuel subsidies.

The vast demand for biodiesel, particularly in the European Union, where half the vehicle fleet runs on diesel, helped drive a rapid expansion of palm-oil plantations in the tropics, especially in Indonesia and Malaysia, to meet the increased demand for vegetable-oil exports. Oil palms produce a higher-quality, higher-yielding, and far cheaper biodiesel feedstock than virtually any other oil seed. Though palm oil is still a relatively minor source of biofuels, its nonfood use rose 450 percent between 1998 and 2010, while production more than doubled.

It is basic economics that when demand rises, prices rise until supplies catch up to meet demand, and that is just what happened. World food prices had been relatively low and stable for decades, bottoming out in 2000. And since grain is perishable and expensive to store, many nations had been whittling down their stockpiles, leading to the lowest grain reserves since the early 1970s—the last time food security was a global concern. But starting about 2002, as demand for corn and oil seeds rose from all the new ethanol and biodiesel plants cropping up around the globe, as sure as clockwork prices for those commodities started to rise.

American farmers had planted an average of about 76 million acres of corn annually during the low corn prices of the 1990s. But with the new ethanol-driven demand, the acres planted with corn in the United States began to rise, jumping 23 percent in 2007 alone. Farmers grew corn at the expense of soybeans, which covered 16 percent fewer acres that year. The resulting smaller soybean crop drove a 75 percent rise in soybean prices between April 2007 and April 2008. Since soy oil competes with other vegetable oils on the world market, the rise in soybean prices triggered a rise in other oil seed prices. Wheat and oil seed crops thrive in the same climate and are grown in many of the same areas around the world. Biodiesel crops like rapeseed and sunflower displaced wheat throughout the European Union, as well as in other wheat-exporting countries, like Argentina, Canada, Russia, and Ukraine. According to the World Bank, the eight largest wheat-exporting countries expanded their planted area of rapeseed and sunflower by 36 percent between 2001 and 2007, while wheat hectares fell by 1 percent. It doesn't sound like much, but it reduced global stocks by more than half and sent wheat prices through the roof.

Rice was another victim. Half of the world's people depend on rice to provide the bulk of their daily calories, and most countries consider it far too important a food crop to brew into biofuel. Yet the rising cost of other grains, particularly wheat, drove up the price of rice as well, which nearly tripled in the spring of 2008, despite minor changes in global production and ending stocks. Rising rice prices led to a desperate run on the global rice market by major importing countries like the Philippines, and bans on exports by major rice producers like Thailand, Vietnam, and India, to protect their citizens from the global price surge.

With spiraling food prices wreaking havoc around the world, biofuel producers were quick to point the finger elsewhere. They blamed high oil prices, droughts in Australia and Russia, and increased demand for pig and chicken feed in China and India. They blamed a weakening dollar and speculators in commodity markets, which saw a fourfold increase in futures contracts for wheat on the Chicago Board of Trade between 2002 and 2006. While all of these factors played a role, most economists placed the blame squarely on biofuels. The International Monetary Fund (IMF) estimated that biofuels contributed to nearly half the 43 percent increase in the price of food from 2006 to 2007. The World Bank, the International Food Policy Research Institute (IFPRI), and two former chief economists with the USDA concluded that biofuels drove up the price of corn and soybeans by 40–70 percent. The IFPRI report went even further, finding that biofuels contributed to more than a 20 percent rise in the price of rice and wheat—which aren't even used for biofuels.

American consumers spend about 14 percent of their daily income on food, most of which is highly processed, meaning that raw materials like corn or soy account for only pennies of the total cost. In the United States and Europe, the price rise was barely noticed. But in the developing world, where people spend on average 50–70 percent of their income on food, the price spike of 2007–8 drove an estimated 100 million more people into poverty.

Even as the industry hit the blend wall in 2012, effectively capping production, its pressure on corn prices remained. That year the US Midwest experienced its worst drought in more than half a century, and the oil and livestock industries lobbied the EPA to waive the ethanol mandate to reduce prices. Noted agricultural economist Bruce Babcock at Iowa State calculated that such a move would have a minor impact, but he added, "The results of this analysis cannot be interpreted as concluding that ethanol production has no impact on corn prices. If US ethanol consumption were somehow banned, then US corn prices would drop to an average of $2.67 per bushel." That month, corn hit a record high of $8.28 per bushel.

IN THAT REGARD, biofuels have been the best thing to happen to US corn farmers since the farm program began in the 1930s. During the height of the biofuel boom, I drove across the Cornhusker State to find out exactly how good things had gotten in a land where two-dollar corn had been the rule for nearly the last three decades. Nebraska is the quintessential flyover state—the land of crop circles and endless fields glimpsed from a 747 window. But down on the ground, with the frost heaves of the highway thumping the tires, it's another world altogether. The land undulates out of Lincoln like a gentle ground swell on the ocean, as uniform rows of corn and soybeans rise above deep soils shoved in place by glaciers. Between the towns, neat, white farmhouses dot the 1-square-mile sections, with the transit-straight, gravel section roads dividing the land into uniform 640-acre blocks. It's a place kids leave in droves, seeking better jobs and opportunities in the cities than on the land their pioneer forefathers carved out of the desolate prairie.

It's also the kind of place where people still take time to chat with strangers. That's how I met Roger Harders, a farmer in his 60s, polishing off a piece of pie at the Wigwam Café in Wahoo. Harders has raised corn, soybeans, alfalfa, oats, and cattle for most of his life. "This is the first year I planted all corn, no beans,"

Harders said. "We also feed cattle and it's expensive. It's going to be a huge adjustment for the cattlemen. I think ethanol is going to raise beef prices, definitely, especially premium cuts. And cattlemen will end up cutting back on the numbers they produce."

I asked Harders if the higher corn prices had convinced him to upgrade any of his equipment, since farmers will typically nurse along an old tractor or combine until commodity prices rise. Harders said he was considering buying a new skid loader and a new manure spreader, as well as replacing an older tractor with a newer used one, though it still would cost $60,000. And he had bought a new truck, though he said he was going to do that anyway.

You could definitely feel the ethanol love over at the Virgil Implement company in Wahoo, where owner Gary Rasmussen sold Case/International Harvester equipment. "Combine sales are up 50 percent, tractors up 10 percent," said Rasmussen. "Any time you get a surge in commodity markets, [farmers] see a brighter future; they have more confidence. Combine sales were tremendous in December and January. We sold 10 new ones at about $270,000 a piece. I don't know how they could invest the money in these things if they don't think the prices are here to stay. I always pull up to the ethanol pump, just 'cause it's going to help my business somewhere down the line."

But one gray-haired farmer sitting on a stool at the parts counter was skeptical. "I think the price will remain high for a year or two and then, watch your pants," said Duane "Woody" Anderson, who's weathered blue eyes had seen more than one boom come and go. "Now, I don't know anything. But all the [fuel] use is still on the coasts and there's no pipeline that I know of. Sure it gives me more market for my corn. But a lot of guys are hanging their butts out. You get to sell alcohol to the oil company. I'm then in the oil business." Anderson shoved back his cap and slowly took a sip of black coffee from a Styrofoam cup. "There's light cattle coming, and light hogs coming too," he said. "Greed is quite a thing. What a revelation."

The ethanol boom has had other unintended impacts as well. Since 2006, US farmers have plowed up and planted 10 million acres of highly erodible land and wetlands that had been enrolled in the federal Conservation Reserve Program, as well as an additional million acres of virgin prairie that had never been farmed. They then dumped more than a billion additional pounds of nitrogen fertilizer onto US farmland to reap the biofuel windfall. Nitrogen levels in the two rivers that supply water to Des Moines, Iowa, were so high during the summer of 2013 that city officials had to invoke water conservation measures so that the treatment plant could keep up with the extra load.

The biofuel boom is a global phenomenon. Brazil has used alcohol fuel made from sugarcane since the 1970s, and most of its cars can run on pure alcohol, pure gasoline, or any mixture in between. Unlike ethanol made from corn, sugarcane ethanol has an energy balance that approaches a whopping nine to one— that's nine units of energy produced for every one unit of energy required to make it. Several life-cycle analyses have pegged sugarcane ethanol's greenhouse gas reduction at nearly 80 percent over gasoline, with an additional 12 percent savings possible. From 2003 to 2008, cane fields expanded by more than 50 percent to meet biofuel demand. Researchers estimate that the area planted in sugarcane will double again by 2018.

Like corn ethanol, sugarcane ethanol has a downside as well, including water pollution, air pollution from the common practice of burning cane before harvest, and labor abuses that stem from child labor to outright slavery on plantations that are still harvested by hand. Despite sugarcane's bitter past and present abuses, it continues to spread across central and southern Brazil, expanding from the state of São Paulo to neighboring Goiás, Minas Gerais,

Paraná, and Mato Grosso do Sul. Much of the expansion is now taking place in the Cerrado, Brazil's vast and diverse savanna ecosystem, where cane typically displaces cattle ranches and other crops. From 2005 to 2008, more than 10,000 square kilometers of the Cerrado morphed into sugarcane fields.

Such land-use change—particularly from native forest or prairie to agricultural field—is one of the largest sources of human-caused greenhouse gas emissions, accounting for roughly a tenth of all annual global emissions. Though the rate of deforestation has slowed in recent years, between 1990 and 2010 the gold rush for timber, cattle, soybeans, and sugar led directly or indirectly to the destruction of 347,000 square kilometers of Amazon rain forest—an area nearly the size of Germany. Elsewhere in the tropics, oil-palm plantations have expanded more than 40 percent since 1990, covering some 15 million hectares by 2011. Most of that expansion occurred in Malaysia and Indonesia, where more than 60 percent of the new plantations created between 1990 and 2005 were carved out of intact rain forest. In Malaysia alone, nearly a million hectares of oil palms were planted on newly cleared peat swamps. The conversion resulted in massive releases of carbon into the atmosphere and the loss of major annual carbon sinks.

The vast majority of the biofuel boom has been driven by large corporate farms, as well as major global corporations that can afford the heavy start-up costs. In 2000, foreign companies owned only 1 percent of Brazil's ethanol industry. By 2010 they owned a quarter. There were nine mergers in early 2010 alone, and the investors were some of the largest players in the energy, agricultural, and financial worlds. They included Royal Dutch Shell, the Carlyle Group, Goldman Sachs, Abengoa, the Soros group, ADM, British Petroleum, Bunge, Cargill, Louis Dreyfus, Mitsubishi, and Mitsui, among many other lesser-known multinationals from China, India, Switzerland, and Norway. When Shell merged

with the Cosan group—the largest ethanol producer in Brazil—
the press release promised that the deal would create "a river of
ethanol flowing from Brazilian plantations to forecourts around
the world."

All over Latin America, Africa, and Asia—particularly in nations
where land rights are fungible and controlled by the government—
vast landholdings are being sold or leased to large corporate agribusinesses to grow biofuel feedstocks, often without consulting or
including the people who actually live on the land and consider it
their own. In 2011, armed military and paramilitary forces in Guatemala violently evicted some 300 families of indigenous Mayan
peasants from disputed land they'd lived on for three years to
make way for sugarcane and oil-palm plantations to supply US and
European drivers with biofuels. The same year, some 427 families
in Kenya's wildlife-rich Tana delta were driven off lands they'd
lived on for 25 years to make way for a sugarcane plantation.

Perhaps the biggest impact of biofuels on the global food crisis,
however, is that the price of basic food grains is now inextricably
wedded to the price of oil. The relationship has existed to a certain
extent since farmers began ditching their mules for tractors and
shipping their harvest on trucks. Corn, wheat, and rice farmers
also use a lot of nitrogen fertilizer, the price of which fluctuates in
lockstep with natural gas. But for most of the twentieth century
the impact of oil prices on food prices was minor. From 1960 to
2008, the World Bank found that every 10 percent rise in energy
prices caused a 5.5 percent rise in fertilizer prices and a 2.7 percent rise in food prices. But when oil rose to more than $80 per
barrel—the break-even point for many biofuels, even without government subsidies—the price of oil and agricultural commodities
rose in tandem.

In 2011 the world's farmers harvested yet another record-
breaking grain crop, which should have brought food prices down.
Instead commodity prices broke records as well. At the G20 meet-

ing of the world's wealthiest countries, the dignitaries listened to a report they had commissioned from all the major organizations that deal with food and the economy. In the report, the FAO, World Bank, IMF, WTO, and IFPRI, among several others, blamed biofuels and government mandates for linking food prices to high and volatile oil prices. They also pointed to another, less visible, culprit that had the potential to drive food prices even higher: "financial investment in commodity baskets."

Unlike the grocery basket that you fill with food for your family, a "commodity basket" is a new financial instrument created by stockbrokers for pure speculation in commodities, including the wheat, rice, and corn that provide more than 80 percent of the world's calories. Just as biofuels were taking off in 2000, Congress passed the Commodity Futures Modernization Act, significantly reducing government controls of the commodity market that had been in place since the 1930s to prevent another speculation-driven disaster like the crash of 1929. Among other things, the Modernization Act exempted from regulation a new type of privately traded, purely financial instrument known as over-the-counter (OTC) derivatives. As financial markets, mortgage markets, and credit markets started tumbling in 2007, large hedge funds, institutional investors, and investment banks sought profits elsewhere, so they moved into commodities en masse. The trade in agricultural futures rose by a third in 2007, while the value of OTC commodity derivatives more than doubled between 2005 and 2007. In the six months between October 2007 and March 2008, the number of contracts on the publicly traded Chicago Mercantile Exchange grew by 65 percent without any increase in actual production. One study showed that 70 percent of the wheat contracts on the Chicago Board of Trade between 2006 and 2010 were controlled not by typical traders or wheat processors, but by large financial speculators.

In the five years leading up to 2012, commodity investments by large institutional investors rose from $65 billion to $126 billion. In the spring of 2012, superspeculators like Goldman Sachs, Morgan Stanley, and Barclays Capital reportedly owned 61 percent of the futures market for all the world's wheat. While millions of children were skipping meals because the price of basic food grains had risen out of reach, the futures traders on Wall Street and in London were making incredible profits. According to London's *Independent* newspaper, Goldman Sachs, now the largest agricultural commodity trader in the world, made nearly a billion dollars on its agricultural investments in 2009, while Barclays Capital, the third-largest trader of food futures, made more than half a billion dollars in 2010 from its speculation on food. The same year, world wheat prices doubled between June and December, even though world wheat supplies remained stable.

In late 2011 the New England Complex Systems Institute (NECSI) released a study confirming what many food experts suspected but had difficulty proving: that speculators were driving up world food prices. The independent think tank and research group is made up of professors from Harvard, MIT, Yale, Brandeis, Duke, Tufts, and several other leading US universities who apply hard mathematical science to complex social, economic, and biological issues. Some of the group's top researchers turned their attention to food prices and developed a sophisticated, dynamic mathematical model of the food price fluctuations since 2004. After modeling every possible cause and effect for the rising food prices—the drought in Australia, the increased consumption of the growing middle class in China and India, the weak dollar, biofuels, and speculation—they came to a rather astounding conclusion: the blame for the food price crises of 2008 and 2011 and the continued record-high prices are due almost entirely to the rapid growth of corn ethanol and the machinations of large institutional speculators. Consider it a new global food tax levied on the world's poorest people by the biofuel industry and the biggest banks on the planet.

The paper's authors called for the immediate restoration of traditional regulation of the commodity market under the Commodity Exchange Act of 1936, as well as a significant decrease in the production of corn ethanol, to relieve the distress of vulnerable populations around the world.

Surprisingly, Congress had already heard this—in 2008. Michael W. Masters, a hedge fund manager from Connecticut, took the unusual step of voluntarily ratting out his fellow speculators. In his initial testimony to the US Senate's Committee on Homeland Security and Governmental Affairs, Masters said, "You have asked the question 'Are Institutional Investors contributing to food and energy price inflation?' And my unequivocal answer is 'yes.' In this testimony I will explain that Institutional Investors are one of [the], if not the primary, factors affecting commodities prices today." The size of these investors, he noted, was staggering. In mid-2008 they had already stockpiled more than 2 billion bushels of corn futures and 1.3 billion bushels of wheat—enough to supply US demand for wheat products for two years. Masters concluded his testimony with an ominous warning:

> There are hundreds of billions of investment dollars poised to enter the commodities futures markets at this very moment. If immediate action is not taken, food and energy prices will rise higher still. This could have catastrophic economic effects on millions of already stressed U.S. consumers. It literally could mean starvation for millions of the world's poor.

By 2011, speculation in agricultural commodities had become so egregious that some 450 economists from around the globe called on the G20 nations to reregulate commodity markets. The Obama administration introduced regulatory reforms to over-the-counter derivatives and commodity trading as part of the Dodd-Frank Wall Street Reform and Consumer Protection Act of 2010, including position limits on agricultural commodities held by large financial institutions to prevent excessive speculation in the OTC market, which is now eight times larger in value than the regulated markets. Even though the federal Commodity Futures Trading Commission and its predecessor, the Commodity Exchange Commission, have limited speculative commodity positions since 1938, Wall Street firms have launched ferocious lobbying and legal challenges against the new limits. Four years after Dodd-Frank was passed, the regulations remain in legal limbo.

The NECSI researchers released a companion paper to their modeling work on speculators and biofuels that was even more concerning. In it they compared the dramatic rise of the FAO's Food Price Index to rising violence and social unrest in the developing world from 2004 to 2011. The first big spike, in 2008, coincided with the food riots that occurred in more than 20 nations; the second big spike, in 2011, coincided perfectly with the Arab Spring uprisings that led to violent protests in more than a dozen countries; toppled long-standing regimes in Tunisia, Egypt, and Libya; and plunged Syria into civil war. Uncannily, the researchers had submitted an earlier paper to the US government essentially predicting social unrest in the Middle East just four days before the first protesters began marching in Tunisia, waving baguettes of increasingly dear bread in their hands.

A farmer gets caught in a dust storm near Felt, Oklahoma, during the drought of 2013, the worst since the 1930s "Dust Bowl."

CHAPTER 7

The Gauntlet

When acre has been added to acre till all the fertile land is occupied, the yearly increase of food must depend upon the amelioration of the land already in possession. This is a stream which, from the nature of all soils, instead of increasing, must be gradually diminishing.—*T. R. Malthus, 1798*

On September 12, 2009, the man called "the greatest hunger fighter of our time" quietly passed away at his modest home in Dallas, Texas. Over the course of his long and productive life, the famous agronomist was said to have saved nearly a billion people from starvation, and raised the living standards of hundreds of millions more small farmers with his miraculous seeds. Norman Borlaug was 95.

In some ways the green revolution died with him. With a glut of grain depressing farm prices in the 1980s, funding for the global agricultural research centers that he'd helped create plummeted, leading to hundreds of layoffs and even the shuttering of one institution. Funding for many national crop-breeding programs was slashed as well, limiting local governments' abilities to adapt seeds to local conditions and distribute them to farmers. In an attempt to cut ballooning debt in developing nations, the World Bank and the International Monetary Fund forced poor countries to slash agricultural extension programs and subsidies, even though the banks' funding nations spent billions subsidizing their own farmers each year.

Borlaug had hoped that new genetic technology would create seeds with broad benefits for poor farmers. Instead, multinational seed and chemical companies created expensive, patented seeds targeted primarily to wealthy farmers of the West. Even wheat rust, the scourge that Borlaug conquered early in his career, has resurfaced. A virulent new strain of the disease appeared in Uganda in 1999 that overwhelmed the *Sr2* complex of genes developed by Borlaug and colleagues. Almost overnight the world's wheat became vulnerable to the newly named UG99 virus. The windborne spores quickly hopscotched to Kenya and Ethiopia and then crossed the Red Sea to Yemen and Iran. Favorable winds or an unwitting freighter or airliner is all that's needed to carry the virus to the world's great breadbaskets: Pakistan, Afghanistan, India, Kazakhstan, Uzbekistan, Ukraine, and even Australia, South America, and the United States. When UG99 was discovered, the number of USDA researchers working on wheat rust resistance had dwindled to two.

At Borlaug's death, about 13 percent of the world's people still suffered from chronic hunger or malnutrition—far less than the 23 percent in 1970, when Borlaug won the Nobel Prize. But the sheer number of people had doubled thanks to population growth. The number of hungry people on the planet today is roughly the same as the entire world population of 1798, when Malthus first penned his thoughts on the relationship between the two.

The distribution of the world's food, energy, and wealth remains as skewed as ever. It's no coincidence that in the years preceding the Arab Spring, food security deteriorated in nearly every Arab country. A study by IFPRI in 2011 found 35 percent of the people in the Arab world (some 120 million) dissatisfied with their standard of living, with the highest proportion in Iraq and Yemen, countries with long-lasting civil conflict.

So even the geopolitical and national security concerns that drove the first green revolution haven't gone away, while the environmental damage to soil, water, forests, and climate has only grown more dire. Despite the great leap in technology and productivity during Borlaug's career, crop yields are now experiencing the classic pressure of diminishing returns that Malthus articulated two centuries ago and that Borlaug himself predicted, while demand for grain from population growth, meat-heavy diets, and biofuels is keeping food prices near record levels and spawning political instability in the poorest countries on the planet.

If we remain on our current path, the challenge before us is stark: In order to stave off widespread hunger, misery, violence, and ecological destruction in much of the globe, we will need to grow nearly twice the amount of food that we currently produce by 2050. And that doesn't even include the amount we'll need to feed an additional 1.5 billion people by century's end.

Most agricultural researchers are down-to-earth pragmatists by nature, and the least likely scientists to be alarmists. Yet when I've asked them how we can accomplish that seemingly impossible task, they strain to come up with answers, and I hear grave concern in their voices. One respected plant breeder I interviewed told me privately that it would be political suicide for the FAO, IFPRI, the

World Bank, or any international organization working on global food issues to come right out and say that we are headed for an agricultural Armageddon. No one wants to be lampooned as the next naysaying Malthus. Yet even these typically staid groups are ringing alarm bells as loud as they can muster. The hard reality is that unless we radically alter the way we live, eat, and farm—or are blessed with a technological miracle as beneficial as Borlaug's short-stemmed wheat—it's hard to see how we will be able to feed more than 9 billion people by 2050 without adding hundreds of millions to the burgeoning ranks of the hungry or plowing up every acre of potentially arable rain forest, savannah, and prairie in a desperate attempt to make our agricultural ends meet.

Here are a few of the high hurdles we must leap in the race to feed an estimated 9.7 billion people by 2050.

Production and Arable Land

Three things contribute to annual per capita food production: (1) the total area harvested; (2) the total annual yield per hectare; and (3) the total population. Leaving aside population for the moment, the first two factors present enormous challenges. Unless we start growing substantial amounts of our food in skyscraper greenhouses—a heady but unlikely scenario, given the costs of pumping all that water and burning all those grow lights— humanity will continue to depend on the thin layer of soil that is warm enough, fertile enough, and wet enough to grow the crops and pasturelands we need. In 2012, the arable land devoted to agriculture amounted to more than 4 billion hectares (about a third, or 1.5 billion hectares, for annual and permanent crops and two-thirds for pastures). That's about 38 percent of the ice-free land on Earth. Agriculture is, by far, the largest human footprint on the planet. Another 4.1 billion hectares is covered in forests—some of which could potentially grow crops. Though deforestation rates have slowed since the high of 16 million hectares a year during the 1990s, an estimated 13 million hectares a year still fell to the chainsaw and the torch in the first decade of the twenty-first century to make room for more pastures or fields.

Current forests remain standing for good reason, though. After 10,000 years of agriculture, humans have already converted forests to fields on the best potential farmland. Those that remain are typically too wet, too dry, or too steep to farm efficiently. Unlike rich prairie or river delta soils, the remaining forest soils are also typically poor, requiring a lot of lime and fertilizer to produce decent yields. We've also protected many of our forests for parks, wildlife habitat, and the services they provide, such as flood control, water purification, and carbon sequestration. Ending deforestation is also one of the primary ways to reduce our carbon emissions.

Some estimates suggest that global agricultural lands could expand by about 5 percent by midcentury, though much of that land is suitable for growing only certain crops, while endemic diseases like sleeping sickness, malaria, and rinderpest, combined with relative infertility, lack of irrigation, and infrastructure, afflict large swaths of potentially arable land as well. As a result, 90 percent of the increase in food over the next 40 years will have to come from land already in cultivation.

To make matters worse, some countries with the highest yields, such as China and the United States, are losing farmland at a rate of about a million hectares a year to roads, housing tracts, office towers, and strip malls. Already the amount of agricultural land per person is falling; having reached 0.26 hectare per person in 2000, it is expected to fall to 0.15 hectare per person by midcentury. That's just over a third of an acre, or slightly larger than the average lot size for a single-family home sold in the United States in 2013. Meanwhile, 8 percent of the world's available farmland will be tied up growing crops for biofuels, the production of which experts from the FAO and the OECD (Organisation for Economic Co-operation and Development) expect will double between 2009 and 2019.

If you shift your mental image of the world's farmland from two dimensions to three, the picture worsens. No less a luminary than Plato warned us of the danger of erosion, which he believed was draining away the wealth and prosperity of his beloved Athens. "The rich, soft soil has all run away," the philosopher despaired, "leaving the land nothing but skin and bone." Over the last few decades archaeologists have confirmed Plato's fears, discovering dramatic episodes of soil erosion that coincided with the rise and fall of great civilizations—not only in Greece, but in Rome, the Middle East, and Mesoamerica. Today, the top layer of Earth's surface is quietly draining away to streams, rivers, and the sea, despite well-known farming techniques designed to protect soil, such as terracing, strip tillage, and no-till farming. By some estimates, the amount of farmland lost to development and degradation amounts to about 30 million hectares each year—an area roughly the size of the Philippines.

So if available farmland is a hard wall that farmers will soon reach, the X factor is yield growth. The human ingenuity, technology, innovation, organization, and investment of the green revolution increased total cereal yield by 89 percent from 1961 to 1986. It was a staggering achievement—perhaps the largest increase in grain production since the invention of the tractor. The bad news is that 1986 was the year the world hit "Peak Grain." During the following 20 years, from 1987 to 2007, population grew faster than crop yields and the harvested area actually fell by 2 percent. By 2009, global cereal production had fallen to 350 kilograms per person, 6 percent less than the peak of 1986.

In 2009 Chris Funk (a researcher with the US Geological Survey) and Molly Brown with NASA Goddard Space Flight Center's Biospheric Sciences Laboratory compared three big data sets that affect food security around the world, taking into account world population and numerous projections for rainfall, climate change, and other factors. What they found, from a food perspective, was terrifying.

Given the growing gap between population and production and the limit on arable land, Funk and Brown concluded that the world will soon be essentially in the same place it was in the 1960s before the green revolution took off, when widespread hunger was the norm for much of the developing world. This was when Stanford biologist Paul Ehrlich was predicting global famines and others were suggesting that the United States triage food aid to hungry nations.

"Globally, a return to per capita production of the late 1960s, when per capita production was near 327 kg per person, appears likely," Funk and Brown wrote. "Eastern Asia may return to 1980s production levels (near 300 kg person^{-1}) and Southern Asia to 1960s levels (near 200 kg person^{-1}). Declines in heavily populated Asia could re-expose millions of people to chronic undernourishment. Central and Southern America may experience 18–20% declines in per capita cereal production levels. Eastern and Middle Africa, however, may be affected most, with more than 30% reductions in already low per capita cereal production levels, with Eastern Africa changing from 131 kg per year in 2007 to 84 kg per person per year in 2030."

Their modeling—which makes the optimistic assumption that developing countries will continue significant yield growth—suggests that by 2030, the world will still produce enough grain to maintain our growing population—although at a low level of 1,900 calories per person per day (the average American consumes and wastes nearly twice that much). But the discrepancies between the haves and have-nots will become even starker. Central and southern Asia may produce enough for only 90 percent of their populations, putting hundreds of millions of people at risk. The situation in eastern Africa is even worse, with cereal production expected to provide only enough to feed 44 percent of its population in 2030, leaving 277 million people dependent on food aid or trade.

But yields have not continued to grow as fast as they did during the green revolution. Rice yields in top-producing countries like China and Indonesia haven't budged significantly in more than a decade, nor have wheat yields in India or France. Increasing them would take either an enormous investment in irrigation or some transformational agricultural technology—both deemed highly unlikely by agricultural experts.

In a 2007 interview, legendary rice breeder Peter Jennings, one of the three men credited with developing the "miracle rice" IR8 at the International Rice Research Institute (and the one who was asked by President Lyndon Johnson to stand out of the picture) was frank about what he and his fellow green revolutionaries had accomplished: "We got seduced by how easy it was to produce IR8 and semi-dwarf wheats," Jennings told the interviewer. "Variety is like a nova: it explodes and then it's dead! . . . Everyone talks as if the Green Revolution is a long-term thing. I don't think so. It was a one-semester shot."

Population and Water

You can have the best soils and the most highly capitalized, technologically advanced farmers planting the best seeds in the world. But without water, you don't have squat. Nothing grows without water, save cactus and tumbleweeds. That's why we use 70 percent of the world's available freshwater supplies to irrigate our crops and water our livestock. Most of it is piped from rivers, lakes, and reservoirs, or pumped from groundwater aquifers—the latter amount having tripled since the 1950s. Water tables are falling dramatically in many of the world's grain baskets. According to the World Bank, 15–35 percent of global water withdrawals were unsustainable in 2009—meaning that water is being pumped out faster than it is being replenished. Over the next few decades, groundwater depletion could cripple agriculture around the world.

Since irrigated farmland produces two to three times the yield of rain-fed fields, the pressure to expand irrigation is rising. The United Nations World Water Development Report in 2012 projected an 11 percent increase in irrigation demand and a 70–90 percent increase in total water demand by 2050. There is likely to be intense competition for that water from industry and expanding cities—the OECD estimates a 400 percent increase from industry alone—which can typically pay far more for it than farmers can.

Water is no longer simply an environmental or food security issue, but a national security issue. In early 2012, a report from the US Defense Intelligence Agency concluded that global freshwater demand was likely to rise to more than 40 percent of the estimated sustainable supplies as early as 2030, threatening world food markets and leading to increasing political instability and reduced economic growth. South Asia, the Middle East, and North Africa will be hit the hardest and may lack the water necessary to grow food and produce energy through hydropower. The risks of the water wars are rising, particularly in areas that are already suffering political conflicts, including India, Pakistan, Israel, the Palestinian Territories, Syria, and Iraq. Key water basins at risk include the Brahmaputra shared by India and Bangladesh and the great Amu Darya of central Asia. The latter is an aquatic lifeline for more than 40 million people living in several "stans"—Afghanistan, Kyrgyzstan, Tajikistan, Turkmenistan, and Uzbekistan—before it dries up in the desert that once was the Aral Sea.

In fact, the rate of groundwater depletion in a 2,000-kilometer swath of Pakistan, northern India, and northern Bangladesh has been calculated at a whopping 54,000 cubic meters per year—70 percent higher than during the 1990s. Detected by NASA scientists working with Germany's GRACE satellites, which measure minute differences in Earth's gravity with incredible accuracy, the groundwater depletion under northern India is the largest loss of Earth's mass outside of the melting ice sheets in Greenland and Antarctica.

To put such figures in perspective, in 2012 some 2 billion people lived in areas that suffered from water stress or scarcity. In two decades that number will rise to 3.6 billion, half the current population of Earth.

Which brings us to perhaps the biggest issue of all: population. Predicting our future numbers is an inexact science, depending, as it does, on assumptions about human behavior in the bedroom. Since we are dealing with billions now, small changes in variables such as life span, birth rates, or death rates can have major effects on the number of future births. Yet we are pretty good at making predictions. The most important of these numbers, Malthus discovered, is the fertility rate, the average number of children a woman will have during her childbearing years. Demographers at the United Nations Population Division have been projecting population growth for individual nations, regions, and the world since the 1950s. In all 12 estimates of world population in 2000, all but one were within 4 percentage points of the actual number of people on Earth.

The good news is that the global fertility rate has fallen dramatically since the 1960s—from 6 children per woman to 2.5 in 2012. In most of the developed nations of the world, the fertility rate is at or below the replacement value of 2.1 children per woman. For example, 2 children "replace" 2 parents, eventually creating a stable population, and the 0.1 accounts for infant mortality. If fertility rates fall below replacement value, the population of a nation eventually falls unless it is supplemented by immigration. In countries like Russia (which has a fertility rate of 1.6), the population is slowly imploding.

In much of the developing world, however, the fertility rate is far higher, and in some countries, such as Nigeria (fertility rate 5.56), the population is still exploding. In 2004, UN demographers predicted that the global population would hit 8.9 billion by 2050, peak at 9.2 billion around 2075, and then hover just below 9 billion until 2300. The bad news is that they've since had to revise their numbers upward as some developed countries like the United States have experienced a slight bump in fertility rates during the early twenty-first century (the US fertility rate was 2.12 in 2007) and fertility rates in many developing countries have stubbornly refused to fall. In 2015, the medium-population projection—which tends to be the most accurate—predicted 9.7 billion mouths to feed by 2050, surging to 11.2 billion in 2100 before flattening out. But hitting even those immense numbers depends on a rapid reduction in fertility rates in the least developed countries—an assumption U.N. demographers admit has "significant uncertainty." If all countries had fertility rates even half a child higher, global population would soar to *16.6 billion* by 2100. Such is Malthus's "power of population." Nearly all of the future population growth will be in the developing world, especially South Asia and Africa.

Population Action International, a Washington, DC–based think tank, argues that the UN's current medium-range forecast is far too low. Its studies show that some 215 million women in developing countries want to avoid pregnancies but have no access to modern contraception. These include countries like Yemen, Afghanistan, and many others in sub-Saharan Africa, where women still give birth to more than 5 children, on average, over their lifetime.

Free condoms aren't enough. Contraceptives have been free in

Niger since 2002, yet the country still has the highest fertility rate in the world—more than 7 children per woman. Married men and women surveyed there in 2006 say they actually want between 8 and 12 children. This is in one of the world's poorest countries, where only 15 percent of the land is arable and droughts and famine are commonplace. Nigeria, Africa's most populous country, lies just across Niger's southern border. The population there is expected to grow 150 percent by 2050.

Climate Change

Each of the immense challenges we face from limited arable land, flattening yields, water stress, and population growth will be exacerbated by climate change. According to the 2014 estimate by the Intergovernmental Panel on Climate Change (IPCC), average global temperature will most likely rise between 1°C with strong emissions reductions and 4°C without them by 2100. That's compared to a 0.5°C rise during the twentieth century that is already melting the Arctic ice cap and the world's glaciers.

When scientists first began studying climate change in the late 1970s, most believed that its effect on world agriculture would be neutral. Arable land that became too hot or dry for crops would be offset by land in the colder regions of Russia, China, or Canada that would warm enough to make farming possible. Yield loss from increased heat stress, they figured, would be more than offset by yield boost from the extra CO_2 in the atmosphere—a conclusion that was supported in several laboratory trials.

When researchers at the University of Illinois began moving those experiments into actual fields in the early 2000s and spraying them with CO_2 from tanks to mimic the concentrations expected during the twenty-first century, they hypothesized as much as a 30 percent increase in yields. Instead they got about a 15 percent increase in soybean yield and no increase in corn yield at all. When they subjected the crops to the higher temperatures or increased water stress predicted for the coming century, both crops suffered yield declines.

"One of the things that we're starting to believe is that the positives of CO_2 are unlikely to outweigh the negatives of the other factors," Andrew D. B. Leakey, one of the researchers, told the *New York Times*.

When the Illinois researchers projected their findings out to 2100, basing their extrapolation on the temperature increase expected for the United States, they predicted that the average weighted yields of these two crops would fall by 30–46 percent by the end of the century under even the IPCC's slowest warming scenario (known as scenario B1), while plummeting 63–82 percent under the fastest warming scenario (A1F1). Outside of the United States, the impacts on agriculture could be catastrophic. The International Food Policy Research Institute has projected that yields might fall by 13–50 percent by 2050, depending on the crop, jeopardizing the food security of more than 2 billion people in Asia alone.

Rising water levels pose another problem. Most of the major "rice bowls" of Asia are along the winding low-lying river deltas of the Ganges, Brahmaputra, Irrawaddy, and Mekong Rivers, which lie less than 2 meters above sea level. Chronically malnourished Bangladesh, one of the most densely populated nations on the planet, could lose 18 percent of its surface area to sea level rise. According to the IPCC, a 1-meter rise in sea level could displace 24 million people in Bangladesh, India, and Indonesia and submerge 102,000 square kilometers of land from Vietnam to India—a swath the size of South Korea.

In 2013, the IPCC calculated that global temperature rise could be contained to the relatively manageable 2°C only if the total amount of carbon released was kept below a trillion tons. At our current rate of energy consumption, we are on track to burn the trillionth ton in 2040.

Wu Wei

So here we are. Fifty years after the greatest jump in agricultural productivity the world has ever seen, we seem to be right back where we started. The world's farmers face a Sisyphean, if not Herculean, task: to double grain, meat, and biofuel production on fewer acres with fewer farmers, less water, higher temperatures, and more frequent droughts, floods, and heat waves. And they have to do it without destroying the forests, oceans, soils, pollinators, or climate on which all life depends. It is the biggest collective hurdle humanity has ever faced.

At this point in our story I'm reminded of one of my favorite *Far Side* cartoons by Gary Larson. Two men are fishing in a rowboat on a lake beside a large, hourglass-shaped cooling tower spewing forth an enormous mushroom cloud. As one of the fishermen looks at the ominous symbol of impending doom, the other says, "I tell you what this means, Verne. To heck with the limit!"

If history teaches us anything, it's that humanity tends to be reactive, not proactive. No less a luminary than John Maynard Keynes observed in 1939, "The idea of the future being different from the present is so repugnant to our conventional modes of thought and behavior that we, most of us, offer a great resistance to acting on it in practice." It's far easier for us to deny that these looming crises exist and continue our old ways—cutting forests, pumping water, and burning fossil fuels—until Earth looks like Mars.

There is another way, of course—an old way that has been around for centuries. Some 2,500 years ago an elderly Chinese philosopher named Laozi ("old master") purportedly saw civilization crumbling around him as the Zhou dynasty devolved into what is now known as the "Warring States period." Although he was an honored member of the court, Laozi decided to abandon his position, leave the kingdom, and embrace the simple life of a monk. As he reached the final gate of the realm, the gatekeeper begged him to write down his teachings on paper, so that his wisdom would not be lost with him in the wilderness.

The result, so the story goes, was the *Daodejing*, "The Classic of the Way and Virtue." It is the most translated work of literature after the Bible and is similarly revered more in theory than in actual practice. Laozi blames humanity's unnatural desires for creating competition and strife. Over the centuries the text was modified by Laozi's followers and he was deified, so, like Christ, no one knows whether Laozi actually existed. But the wisdom in the little book seems relevant to the challenges we now face.

The *Daodejing* purports to teach us how to live in harmony and balance with each other and the world through *ziran*, the principles inherent in nature. We achieve *ziran* through the process of *wu wei*, which is often translated as "doing nothing." The way Laozi explained it, however, is a little more complex: "lack acting and yet lack 'don't act.'" The great symbol of Daoism is water, a substance that is soft and yielding, yet so powerful it can erode mountains or carve great valleys. Like the raindrop that forms the stream that feeds the river that pours into the vast depths of the ocean, *wu wei* seems to embody the more simple modern saying "Go with the flow."

I learned this concept not from a Daoist monk, but from a stocky mechanical engineer in Arizona who had likely never heard of the *Daodejing*. Ray Hobbs was a 60-ish, potbellied employee of the electric utility company Arizona Public Service (APS) with a pencil-thin mustache, an Appalachian drawl, and an honest-to-

God pocket protector full of pens. He'd spent most of his career building coal-fired power plants to meet the exploding energy needs of the fastest-growing state in the nation. I met him in 2005 while researching a story on biofuels. He was shepherding a demonstration project with some MIT engineers who were trying to make biofuel from algae. The group was siphoning off CO_2 from one of APS's gas turbine power plants. If the scientists could figure out how to grow and harvest algae economically on an industrial scale, they could cut the carbon emissions of the plant in half while pumping out ultragreen biodiesel and ethanol for sale.

I pegged Hobbs as one of those old-school engineers who think climate change is a joke. I couldn't have been more wrong. He was a thoughtful, reflective man with an unbridled passion for alternative transportation fuels. He had a fleet of experimental vehicles from small, futuristic electric golf carts to big hydrogen-powered buses. He'd recently opened the first commercial hydrogen refueling station in the country for the few experimental hydrogen vehicles roaming around Phoenix. He was probably one of the few native West Virginians who believed coal wasn't God's gift to America, and he had his hands firmly on the throat of the global energy conundrum.

"What we have to start looking at in this country is the true cost of fuel," he told me on our long ride out to the Redhawk power plant, past rows and rows of brand-new adobe tract homes stretching to the horizon. "Including health and environmental costs. We want energy cheap because we want to throw it away. That is absolutely unsustainable, and that is our downfall as a country. We've developed unbelievably wasteful habits. In typical urban traffic you only get 4–5 percent of useful work out of your car's fuel consumption. All we are doing with our current use of fuel is making air pollution and heat while doing very little work. You're throwing 95 percent of your money away. It's insane. How can we sustain ourselves as a nation with such a disrespect for our natural resources?"

Hobbs was an enthusiastic supporter of algae, which he called "God's Clydesdale," the workhorse of the natural world that could double its mass in 12–24 hours. The secret, he believed, was not in the science or the mechanics of the bioreactors growing the algae, but in the billion or so years of evolution that had endowed algae with their phenomenal ability to reproduce.

Hobbs's true passion, however, was electric cars—especially electric racing cars. He had a picture on his desk of a stickered-up Saturn coupe he'd built and raced that would do 0–60 miles per hour in less than 5 seconds—comparable to the orange Lamborghini I'd seen in Beijing. He'd started electric-car clubs in the Phoenix high schools, helping students build and race their cars, and had convinced APS to sponsor larger races at the Phoenix motor speedway. "It was utterly quiet, like watching NASCAR with the mute button on," Hobbs said. "All you could hear were the tires chirping through the turns." He even attempted to break the world electric-car speed record at Bonneville with a customized British Lola—a carbon-fiber cockpit car powered by $70,000 worth of military-grade nickel-metal-hydride helicopter batteries, theoretically capable of speeds topping 200 miles per hour. Unfortunately, a short circuit during a test drive fried the batteries (but luckily not the driver), and Hobbs had to pull the plug on the project. The molten carcass sat in his garage covered by a tarp—too painful for him to view. As he was showing me the remains, however, he said without much fanfare that in 10 years we'll all be driving electric cars.

He explained it this way. The average American drives 30–40 miles a day. There are electric cars on the market that go 60–120 miles on the electric equivalent of 1 gallon of gas—which APS sells for about 99 cents. And we can make that electricity out of virtually anything: coal, oil, gas, nuclear, solar, wind, hydropower, geothermal, biomass, methane, landfill waste, wave energy—you name it. And there is a supply line attached to every house. "It's downstream," Hobbs said with a grin. "Going with the flow."

There is no single solution to the looming food crisis. Whether we are compelled by Laozi's admonition to lead simpler, more natural lives or by Malthus's "goad of necessity," or Adam Smith's enlightened self-interest, we're going to need every tool and technique we can find to put food on the world's table. The following chapters explore a few paths that might help us along the way.

PART II

Giant Japanese scallops thrive on waste from nearby pens of sablefish at an integrated multitrophic aquaculture farm off Vancouver Island.

CHAPTER 8

The Blue Revolution

With Earth's burgeoning human population to feed we must
turn to the sea with new understanding and new technology.
We need to farm it as we farm the land.—*Jacques Cousteau, 1973*

Eight miles off the coast of Panama and 30 feet beneath
the cobalt Caribbean, Brian O'Hanlon lay motionless at
the bottom of a giant circular net pen as if stuck in a
gargantuan spider's web. Above him, 35,000 cobia—sinuous,
remora-like creatures nearly a meter long—swam around the
enclosure creating a slowly spinning vortex of fins and 150 tons
of pulsing, muscular aquatic flesh. Below him, 200 feet of clear
blue water darkened into an abyss plummeting to the sandy
bottom of the Panama shelf.

I watched O'Hanlon and the rotating fish ball from a few meters away, trying—and failing—to slow my heartbeat with steady breaths through my regulator. Each summer my friends and I attempt to catch cobia when they migrate into North Carolina's coastal waters. Though they can grow to more than 100 pounds, any one of the fish in the pen would have been cause for backslapping and beers if we'd been lucky enough to land one. Suddenly a 10-pounder stopped inches from my face mask to give me a steely gray stare. I shooed him back to his piscatory parade envisioning the buttery white fillets that were his destiny to provide.

For O'Hanlon, however, the experience was exactly the opposite—a rare chance to slow down and relax. "I love to just sit in the bottom of the cage and look up, breathing underwater, and see the fish swirling around," the young entrepreneur said after we'd clambered aboard one of his company's bright-orange pangas—the ubiquitous blunt-nosed fishing skiffs of the tropics. "It's like meditation. I used to call Sunday dives 'church' because that's kind of what it felt like."

Those peaceful moments are rare these days for the 33-year-old founder of Open Blue, a company trying to turn Jacques Cousteau's 40-year-old dream into a reality. For the past decade, the native Long Islander has been working to create the first profitable large-scale marine fish farm in deep offshore waters. If successful, he could spark a new era of food production—potentially turning the world's oceans into three-dimensional fields producing millions of tons of highly nutritious aquatic protein. Opening just 2 percent of the oceans outside of the Arctic and the Southern Ocean would be the equivalent of adding 1.6 billion acres of fish pasture—roughly five times the area planted with the major crops in the United States each year.

It has been an epic uphill battle. No one, O'Hanlon says, has yet made money farming the open ocean. The vast majority of the world's current fish farms utilize freshwater ponds, lakes, and rivers. Those that have ventured into the storm-tossed sea, such as the salmon industry, try to anchor their pens in shallow, well-protected coastal waters near shore that are far easier and cheaper to manage. The waves over O'Hanlon's cages top 20 feet at times. His dozen 6,500-square-meter pens are built like giant gyroscopes with heavy galvanized steel spars and Kevlar netting—the same carbon-fiber fabric used in bulletproof vests. Their main defense against battering storms is their ability to submerge below most of the wind and wave energy at the surface.

O'Hanlon's first test farm in Puerto Rico was slammed by two hurricanes, including a category 4 whopper on the Saffir-Simpson scale that passed within 100 miles. One pen dragged a 10,000-pound anchor, but he lost no fish. Still, exposing your life's work, not to mention life's savings, to the ocean's might is not for the faint of heart or the shallow of pocket.

Open Blue is one of a handful of highly capitalized, high-risk ventures trying to expand into the aquaculture frontier. But even run-of-the-mill, mom-and-pop fish farms have been growing for decades. In fact, aquaculture has been the fastest-growing food production system on the planet since the early 1970s, expanding on average by more than 8 percent each year. In 2013, the world's aquaculturists grew more than 70 million tons (excluding aquatic plants) from 600 different species, ranging from majestic tunas to sea cucumbers that only a Cantonese cook could love. The world now farms more fish than beef. Nearly half of all the fish and shellfish eaten on Earth are now raised on farms rather than caught or collected from the wild, up from barely 4 percent in 1970. Population growth, income growth, and seafood's reputation as a healthier alternative to four-legged livestock are projected to increase the demand for fish and shellfish by another 50 percent by midcentury.

With most wild fish stocks either maxed out or declining, nearly all that additional seafood will come from fish or shellfish farms. It's no wonder that just before he died, the influential economist, author, and business guru Peter Drucker, who gave us the phrase "knowledge worker," suggested that aquaculture, not the Internet, was the more promising investment of the twenty-first century:

> Within the next fifty years fish farming may change us from hunters and gatherers on the seas into "marine pastoralists," just as a similar innovation some 10,000 years ago changed our ancestors from hunters and gatherers on the land into agriculturists and pastoralists. . . . Twenty-five years ago salmon was a delicacy. The typical convention dinner gave a choice between chicken and beef. Today salmon is a commodity, and is the other choice on the convention menu. Most salmon today is not caught at sea or in a river but grown on a fish farm. The same is increasingly true of trout. Soon, apparently, it will be true of a number of other fish. . . . This will no doubt lead to the genetic development of new and different fish, just as the domestication of sheep, cows, and chickens led to the development of new breeds among them.

Humans have been farming fish nearly as long as we've been fishing for them. Carvings on Egyptian tombs dating to about 2500 BC show men harvesting tilapia from a pond. Pliny the Elder, the great Roman naturalist and naval commander of the first century AD, wrote detailed descriptions of Roman fish ponds where nobles raised fish and eels. On the other side of the planet, native Hawaiians were growing mullet and milkfish in ponds and enclosed lagoons called *loko i'a* centuries before the arrival of Captain Cook—helping to sustain a population possibly as large as a million people. The fall of the Hawaiian nobility led to the end of the fish ponds, while overfishing around the islands in modern times has taken its toll. Today Hawaii imports 85–90 percent of its food to feed nearly 1.4 million residents.

China, however, is the undisputed birthplace and modern leviathan of aquaculture, producing nearly 60 percent of the world's farm-raised fish and shellfish. Farmers there have been growing common carp for thousands of years. A Zhou dynasty official named Fan Lai actually wrote the first textbook on the subject in 475 BC, entitled "The Classic of Fish Culture." Eventually the Chinese developed an intricate polyculture of carp, pork, duck, and vegetables on their small farms, feeding the waste from one species to another. By the Three Kingdoms period, about AD 220, farmers had begun adding carp to their flooded rice fields, and they discovered that the omnivorous fish gobbled up insect pests and weeds but left the rice seedlings alone, providing food, free pest control, and fertilizer for their paddies. The system was so simple, sustainable, and robust that farmed carp and rice became pillars of the Chinese diet, helping to support one of the densest populations on Earth.

The ancient polyculture gave way to modern monocultures during the 1970s and 1980s, when Chinese leader Deng Xiaoping encouraged his country's fish farmers to produce more fish and shellfish not only for China, but for the burgeoning and lucrative export market around the world. They did so with capitalistic fervor, and soon China's lakes, rivers, and coastal zones were jam-packed with nets and fish that no longer grazed on grasses or wastes from livestock but were fed commercial fish feeds. During the same time period, USAID helped finance large shrimp farms in Latin America, while, on the other side of the Atlantic, companies in Norway and Scotland were expanding Atlantic salmon aquaculture into their fjords.

Unfortunately, as the new "blue revolution" packed our freezer shelves with relatively cheap shrimp, salmon, and tilapia from Asia, Norway, and Chile, it brought with it many of the environmental warts of the green revolution. During the 1980s, roughly a quarter of Earth's remaining tropical mangroves were bulldozed to build shrimp ponds that now produce a third of the world supply, driven largely by demand from the United States, Europe, and Japan, the world's largest importers of fish. Las Vegas alone consumes some 60,000 pounds of shrimp a day—roughly 150,000 shrimp cocktails—most of which is imported from Asia.

Aquaculture effluent, a nitrogen- and phosphorus-rich cocktail of dead fish, feces, and rotting feed pellets, is now a major water pollutant throughout the region. Algal blooms and dead zones of hypoxic water are commonplace. The world got a close-up view of the problem during the 2008 Beijing Olympics when sailing events in the coastal city of Qingdao were threatened by a 1,200-square-kilometer algal bloom—the largest ever recorded—that was ultimately traced to commercial seaweed farms 200 kilometers away. The area used for seaweed aquaculture had doubled to 23,000 hectares in the five years prior to the games.

To keep pens of densely packed fish or shrimp alive in polluted water, however, fish farmers often resort to feeding them antibiotics and pesticides banned for use in the United States, Europe, and Japan. This practice has raised major food safety concerns in the United States, which imported more than 2.4 million tons of seafood in 2012. In 2006 and 2007, the Food and Drug Administration discovered that 17 percent of the samples taken from Chinese aquaculture imports contained banned substances, including known or suspected carcinogens, including nitrofurans, chloramphenicol, malachite green, and gentian violet. Nor was China alone. Contaminants were found in shrimp, tilapia, basa, grouper, crab, and salmon from nearly all the major exporting countries, including China, Thailand, Taiwan, Vietnam, and Chile. And these are just cursory checks. Less than 2 percent of US seafood imports are inspected, and only a tenth of 1 percent are analyzed for 13 substances of concern. In contrast, the European Union inspects 50 percent of its seafood for 34 potentially dangerous chemicals.

"China preferred polyculture for thousands of years," says Professor Li Sifa, a fish geneticist at Shanghai Oceans University. "But during the last 30–40 years this system was discarded by the people for monocultures and high-volume systems. Before, we used all water surfaces, rivers, lakes, ponds, canals. But no more because of pollution." Li should know. He is the Norman Borlaug of aquaculture in China, having developed a fast-growing breed of tilapia that is now the backbone of China's annual 1.2-million-ton tilapia industry. He says Chinese fish farmers are using fewer pharmaceuticals, and many of the more toxic chemicals are now banned.

Fish farms in Europe and Latin America have had troubles scaling up as well. Salmon farms have been plagued by parasites, fish escapes, and diseases. Some large-scale farms generate as much raw sewage as a small town, polluting nearby waters. Though the industry has improved its environmental footprint over the years, Chilean salmon farms are still recovering from an outbreak of infectious salmon anemia, or "fish flu," that wiped out more than two-thirds of the farmed fish between 2007 and 2010. Recent disease outbreaks have also devastated oyster farms in Europe and shrimp farms in Asia, South America, and Africa—almost ending the shrimp industry in Mozambique in 2011.

The quest to produce a lot of seafood with minimal environmental impact drove O'Hanlon to Panama. A third-generation fish monger from Long Island, O'Hanlon grew up with New York's famed Fulton Fish Market as his playground. His family was among the first to import farmed salmon from Norway into the United States, but the business fell on hard times after the North Atlantic cod fishery collapsed. The lesson was clear to the O'Hanlon clan: the future was in farming fish, particularly the high-value marine species that consumers prefer.

Most aquacultured fish, such as carp, tilapia, trout, and salmonids are born in freshwater and have relatively simple reproductive processes. They are easily bred and produce large eggs that aquaculturists can capture, fertilize and propagate with little trouble. Marine fish, however, are a different story. Their eggs are tiny, and the ocean is such a vast and mysterious place that we still have only a vague idea of where most marine fish spawn, reproduce, or mature.

As a teenager, O'Hanlon raised mutton snapper at his grandparents' house in Long Island until his 700-gallon aquarium burst, turning their basement into a subterranean swimming pool. He then built a larger aquarium in his parents' house and conducted his experiments there until a short in a ventilation fan nearly burned the place down. In 2002 he left Long Island to study under Dr. Daniel Benetti at the University of Miami, a pioneer in breeding marine species for aquaculture, and shortly thereafter O'Hanlon started a small experimental farm in Puerto Rico, growing snapper in offshore net pens. The fish thrived there but grew very slowly. Only after Benetti sent him a few cobia to try did he have his epiphany. Cobia have been a popular sport fish from the Carolinas to the Texas coast for years. Their meat is firm, white, and delicious, and since they have no swim bladders, the organs that most fish use to adjust their buoyancy, they store fat in their tissues to help them stay afloat, giving them as much omega-3 fatty acids as salmon have. Best of all, cobia grow like crazy. After a year in the pens, the cobia weighed 10 pounds each; the same-aged snapper weighed only one.

But try as he might, O'Hanlon couldn't pry a permit out of the numerous agencies in charge of American coastal waters to expand the operation to commercial scale. He finally gave up, bought a bankrupt fish hatchery in Panama and shifted his cobia south. In the clean, deep waters down the coast from Colón, the current is so swift that O'Hanlon's fish never see the same water twice. His pens are stocked at a fraction of the density of a typical salmon farm and thus far haven't required pesticides or antibiotics. They've also tested free of heavy metals that can sometimes accumulate in fish feed. O'Hanlon is so confident in his product that he feeds his cobia to his two-year-old daughter, and his wife ate it regularly when she was pregnant. Since the pens are located in a part of the ocean that is very low in nutrients, the nitrogen and phosphorus in the fish waste is quickly scavenged by phytoplankton and zooplankton in the water. Scientists have yet to detect any pollution from the farm outside the pens.

O'Hanlon still has to feed his fish, however, and the cost of that feed—as well as the high cost of an offshore operation and the lack of an established market for cobia—remain big hurdles to creating a truly economically and environmentally sustainable fish farm. Most of the marine fish we love to eat—tuna, salmon, snapper, cod, mahi-mahi, cobia—are voracious predators, perfectly adapted to hunting and eating smaller fish. Cobia grows so explosively because growth is a survival strategy in a fish-eat-fish world. To keep his cobia thriving and high in omega-3, O'Hanlon feeds his flock Peruvian fish pellets containing roughly 40 percent fish meal and 10 percent fish oil derived from small schooling pelagic fish such as sardines and anchovetas. The rest is largely soy meal.

Although these species grow thick along the Peruvian and Chilean coasts, they are naturally cyclical, and highly vulnerable to overfishing and changes in sea temperatures associated with La Niña and El Niño events. They go through spectacular booms and busts, such as the one that ended the great California sardine industry, as John Steinbeck captured so colorfully in *Cannery Row*. Pacific sardine populations are now plummeting off North America, South America, and Japan.

Fish farms currently consume nearly 70 percent of global fish meal and almost 90 percent of fish oil. Not surprisingly, the price of fish meal tripled between 2002 and 2010. Some fishery experts worry that the rising price will cause a global rush on sardine and anchoveta stocks, while conservationists are concerned that vacuuming the sea of small oily fish will leave precious little for larger predatory fish, seabirds, and marine mammals that depend on them. Yet so hot is the market for fish meal and oil that Norway—the largest producer of farmed salmon in the world—as well as Japan and South Korea, have resumed harvesting tiny krill off the coast of Antarctica, a major food source for penguins, seals, and whales in the region. The fishery was pioneered by the Soviet Union in the 1970s and fell off after the country's collapse. But the take was tiny compared to the estimated krill stock. No one knows how an expanded krill catch will affect Antarctic wildlife.

Since it typically takes several pounds of small forage fish to make 1 pound of marketable salmon or cobia, critics of the industry have likened farming top marine predators to raising lions for food. Aquaculture advocates, on the other hand, argue that modern aquaculture is still a young industry, and vastly more efficient at using small forage fish than it once was. The ratio of pounds of forage fish going in to aquaculture to farmed fish coming out has dropped to nearly a third of what it was 15 years ago—from 3 or 4 to 1, to 1.5 to 1—for major farmed species. That ratio continues to fall as fish farmers shift to more omnivorous species like tilapia and feed them more soybeans, corn, and canola.

Fish wastes are another alternative feed. Every year the Alaskan pollack fishery, one of the better-managed wild fisheries, throws away more than a million tons of fish guts and fins, known as "processing wastes." This material could be recaptured and turned into fish meal, as could processing wastes from farmed species like salmon—more than half of whose total weight is typically discarded as offal. Some feed companies are even extracting the oils from single-celled organisms like algae, the omega-3 factories of the sea, to replace fish oils. The volumes needed by aquaculture and the current cost to extract algal oils remain prohibitively high for most fish farms. But as costs come down, algae may one day completely replace forage fish in aquafeeds, perhaps as a profitable by-product of an algal biofuel industry. Such algal feeds could also prevent bioaccumulated contaminants like DDT, PCBs, and dioxins from reaching farmed fish, according to Roz Naylor, a food policy expert at Stanford University. An even quicker fix, she says, would be to genetically modify rapeseed or soybeans to produce high levels of omega-3 fatty acids for aquafeeds, if consumers will accept it.

"My long-term vision is to go beyond fish and grow macroalgae in the ocean," says O'Hanlon, noting that some kelps can grow a meter a day. "Fish is just a gateway to really farming the oceans. There's no way around it. . . . Fish meal and oils can be one of the greatest sources of renewable energy in the world if properly managed."

Despite the controversy over aquaculture's impact, recent reports by the WorldFish center (part of the Consultative Group on International Agricultural Research, or CGIAR), Conservation International, and the US National Oceanic and Atmospheric Administration (NOAA) have concluded that farming fish produces protein far more efficiently than either terrestrial livestock systems or wild-capture fisheries, and with a much lighter ecological footprint. Because fish are cold-blooded and live in a buoyant environment, they need fewer calories to regulate their body temperature or build strong bones that counteract gravity. To grow a pound of farmed-salmon body mass requires just over a pound of dry feed. Growing a pound of chicken requires nearly 2 pounds of feed, pork nearly 3 pounds, and beef almost 7 pounds. And marine species require no freshwater.

About 3,700 miles north of the sultry waters of Panama, tucked in a remote corner of Vancouver Island, Dr. Stephen Cross of the University of Victoria is developing another novel type of aquaculture that would be instantly recognizable to a Chinese fish farmer of Fan Lai's day. Using a system known as integrated multi-trophic aquaculture (IMTA), Cross has taken a small salmon farm and turned it into the aquatic equivalent of an organic garden—replete with a few pigs and a variety of heirloom vegetables. He feeds only one species—a sleek, hardy native of the North Pacific known as sablefish or black cod. Just down-current of the fish pens he's placed Chinese lantern baskets full of native oysters, scallops, and mussels that feed on the excretions of the fish. Beside the baskets he grows rafts of sweet sugar kelp—the kind used to wrap sushi—that filter the water even further, removing nearly all the remaining nitrates and phosphorus. On the hard sandy bottom 25 meters below the fish pens, he grows native sea cucumbers, ugly echinoderms beloved by Chinese and Japanese diners, that serve as aquatic vacuums sucking up heavier organic matter that the other species miss. He even plans to grow sea urchins to eat the unwanted marine growth on the farm's cages. Though the project is still in the precommercial stage, Cross's ultimate goal is to produce 11 native species and run the entire operation on solar energy. He has applied to be the first certified organic fish farm in Canada.

"The whole concept of moving aquaculture into offshore waters or onto land isn't because we've run out of space [in the coastal zone]," says Cross, who earned his PhD in Scotland looking at contaminants in farmed fish and shellfish and was an environmental consultant to the salmon industry for decades. "It's because of the conflicts between stakeholders." Fish farms—no matter the species—Cross says, are still paying the price for the dark ages of salmon aquaculture in the 1980s, when the young industry was beset with problems. Now, he says, even salmon farms are producing 20–30 times the fish with a fraction of the environmental loading. He's actually designed his system to attach onto existing salmon farms to serve as sort of a giant, moneymaking water filter. Researchers on Canada's Atlantic coast are attempting something similar, integrating filter-feeding mussels and kelp into large commercial salmon farms, where the two species thrive on the salmon's waste.

One evening at the farm's houseboat, which serves as office, lab, feed shed, and student housing, Cross and Nathan Blasco, a graduate student in charge of the kelp, tossed a few sablefish fillets and biscuit-sized scallops into a pan bubbling with garlic, butter, and white wine. We sat in chairs on the dock as the sun set into the fragrant Douglas firs, savoring a meal worthy of a five-star restaurant and watching for bears along the shore. Between bites, Cross waxed on about his fish like a farmer discussing his favorite breed of cow:

"Sablefish typically live in 1,000 meters of water. They are very long-lived. The Japanese caught one that was 95 years old. Adults live in dark, cold water, with very little food, low metabolism, and low growth. But they spend the initial part of their lives near inlets in surface water, snarfing food. They grow really fast until they get to 1 kilo—about 2 pounds. Then they go offshore, and you never see them again. We get 10-gram juveniles, and when they reach 1.3 kilograms their growth flattens. To get to market size at 2.2 kilos is a two-year process. But that will change. It took five years to figure out how to breed them and another five years to figure out what to feed them. But they are hardy as hell. Just a beautiful fish. You walk down my farm and you never see them. They stay low. I've never had to treat them. We just vaccinate the little guys, and that's it. No antibiotics. If you reduce stocking densities, you reduce stress on the fish. They cost more to produce, but stress is reduced, and the potential for disease is reduced."

Even China is going back to its polyculture roots. The area

under rice paddy aquaculture more than doubled between 1985 and 2007, to more than 1.5 million hectares, producing more than a million tons of fish. Researchers have documented significant reductions in pesticide and fertilizer use in the system, which means lower costs for growers. In some areas, many of those fish and bushels of rice are now marketed as organic, bringing the farmers higher income and Chinese consumers greater confidence in the safety of their food.

Perry Rasso of Matunuck, Rhode Island, on the other hand, doesn't feed his aquatic livestock anything at all—and he's got 12 million of them. Rasso is an oyster farmer, a type of aquaculturist that has been lauded by seafood lovers and environmentalists alike. Besides producing a healthy product low in fat and high in omega-3, his 7-acre farm in a small saltwater pond south of Providence is actually cleaning the water, reducing sediments, nutrients, and turbidity.

Rasso, with his Popeye forearms, five o'clock shadow, and fisherman's hoody, looks more like the collegiate wrestler he once was than the greenest guy in aquaculture. He started his farm his freshman year in college at the University of Rhode Island, and he was soon selling his oysters at local farmer's markets, just as the farm-to-table movement began to blossom. "I'd get to the farmer's market, look around, and say what am I doing around all these crunchy people?" Rasso told me. "But then I started making more money, started eating local foods, and you know what? That stuff was good!" Rasso eventually bought a bankrupt restaurant near his farm just for its commercial dock, and he opened the place almost on a whim, serving oysters from his and other local growers' farms in his co-op, as well as local produce he brought back from the farmer's market. The first day he ran out of coffee. Then he ran out of cups. "I figured when it failed, I'd just live here," he quips. Instead, his Matunuck Oyster Bar is now serving 800 people a day in the summer season, with an hour-and-a-half wait for a table.

The oyster farmer hasn't stopped there. He gives farm tours to inner-city kids from Providence who've never seen the ocean, much less tasted a salty oyster straight from the water, and the University of Rhode Island has sent him far and wide to teach aquaculture in places like the Cape Verde Islands, Tanzania, and Sudan. Though not a traditional form of food production in Africa, aquaculture is exploding there, supplying inexpensive protein and jobs in a region where both are in short supply.

WHILE SOME FISH farmers are heading offshore to reduce their impact on the environment, others are going in the exact opposite direction—moving their fish on land. Bill Martin, a silver-haired businessmen from Martinsville, Virginia, is one of them. Within the walls of a nondescript warehouse tucked in the Blue Ridge Mountains next to the local NASCAR speedway, he's built one of the largest indoor fish farms in the world, producing 10,000 pounds of fat white tilapia every day. He raises them in a series of rectangular concrete tanks, each roughly 15 feet wide by 80 feet long and packed so full of fish you could almost walk on their backs. The actual density is 2 pounds of fish per gallon of water, Martin says—akin to having five fish the size of dinner plates in the typical 10-gallon home aquarium. The day I visited, he picked up a bucket of feed pellets—made mostly from soy meal, since tilapia are omnivores—and slung some into a tank. The big white fish boiled to the surface like piranhas, sounding like a rainstorm as they fed.

"Net pens are a total goat rodeo," said Martin in a raspy south-ern drawl. "You've got sea lice, disease, escapement, and death. They had one recently in which a thousand fish died and no one knew why. The best reason they could come up with was something scared them to death! You compare that with a 100 percent con-trolled environment—a situation that is possibly as close to zero impact on the oceans as we can get. Land-based systems are the way to go. Fifty years from now 50–60 percent of the seafood you eat will come from them. My model is the poultry industry. I want to sell everything but the cluck. We've got greater speed of growth, greater efficiency. The difference is our fish are perfectly happy."

"How do you know they're happy?" I asked.

"Generally they show they're not happy by dying," Martin said. "I haven't lost a tank of fish yet."

The indoor fish farm, known as a recirculating aquaculture system (RAS), as well as the fish it produces, does not come cheap or without risk. Martin went bankrupt trying to grow catfish at the farm, which were plagued by disease. Not until he tried the hardy, fast-growing tilapia, which thrives at high densities, was he able to make it work. Even then, Martin's fish cost more per pound than do wild-caught fish, or even fish grown in ponds in China and shipped to the United States. The farm also pulls 500,000 gallons of water each day from the underground aquifer to supply the 3-million-gallon facility. He recycles about 80 percent of the water, and the rest—high in ammonia—goes to the sewage treatment plant. He's currently experimenting with a new filtration system that will recycle 99 percent of the water, and methane generators to burn the fish waste and help offset the high power bills and his carbon footprint. One day, he says, such systems may even be able to use recycled water from sewage plants.

Martin claims that his newest project is even greener. In a new warehouse next to the tilapia farm, he has built long, shallow, fiberglass tanks stacked two stories high. Here his staff is working with scientists at Virginia Tech to develop an oyster and shrimp RAS, in which oysters feed on the shrimp waste and vice versa—creating a minimally fed, closed-loop system in which most, if not all, of the waste products are recycled.

I first met Bill Martin in early 2011. As we spoke that day, a giant flat-screen TV on his office wall beamed live images of the

demonstrations in Cairo's Tahrir Square during Egypt's Arab
Spring uprising. Martin has worked as an aquaculture consultant in Egypt and other developing countries for decades. He had
recently turned down a deal to build a giant tilapia farm outside
of Cairo, in part because the investors—who were close associ-
ates of former president Hosni Mubarak—were only interested
in exporting fish to Europe, not feeding the hungry Egyptians in
their streets.

"In Egypt they have a great saying: Let me live today, kill me
tomorrow," Martin said. "The problem they got is they've got no
concept of tomorrow. But a man is not an animal. We have a moral
obligation to feed people. When we get this system right, with
ozone-cleaned sewage water, we can give them fish every single
day. And once you feed a man every day, then life tomorrow has
some meaning to it. The African coast had one of the greatest fisheries in the world, and it is gone. The African people didn't get any
of it. This system has a conscience. If the UN wanted to put this
in there, it would become a fish-making machine."

Martin sat back in his chair, his eyes fixed on the images of
young Egyptians rioting in the streets of Cairo. For him the issue
transcended politics. "We must do this or we will have a world
in chaos," he said. "This is toppling governments. And you know
who's going to be their best friend at the end of the day? The guy
who puts the food on the table."

A Soviet propaganda poster—likely from the early 1960s—exhorts farm-workers to "Fight to maximize the use of all collective farm and Soviet farm production reserves. Fight for high crops!"

CHAPTER 9

Back in the USSR

When I am dead, bury me
In my beloved Ukraine,
My tomb upon a grave mound high
Amid the spreading plain.
—Taras Shevchenko, 1845

J ustin Bruch, a 35-year-old farmer from Emmetsburg,
Iowa, leaned back and took a deep pull from a tall frosted
mug of beer. With his crew cut, pressed jeans, and crisp
dress shirt, Bruch (pronounced "brook") looked as if he had just
dropped by his favorite honky-tonk on an Emmetsburg Satur-
day night.

But instead of a tall Bud Light, the young farmer was knock-
ing back a Czech brew that had been around since 1516, and
carving mouthwatering slices off a spit-roasted pork knuckle

that could have been prepared by fourteenth-century monks. The outdoor beer garden sat in the shadow of the 700-year-old stone wall of Lviv, the once proud capital of the erstwhile kingdom of Galicia, but subjected over the centuries to Polish, Austrian, Hungarian, Swedish, and Russian-Soviet rule. Known variously as L'vov (by the Soviets), Lwów (by the Polish), and Lemberg (by the Austrians), the medieval city, with its cobblestones and Gothic churches, is perhaps best known as home to writer Leopold von Sacher-Masoch, whose tales of fur-wearing, whip-bearing dominatrices gave rise to the fetish of "masochism." It is an unfortunate legacy for Lviv, now the cultural hotbed of the splintered, self-flagellating nation of Ukraine.

Ukraine is a world away in miles and mentality from the Corn Belt of Iowa, but the two places have a few things in common. First, the homeland of the horse-loving, mercenary Cossacks, among several other ethnic groups, is a slumbering agricultural powerhouse, long renowned as the breadbasket of Europe. About the size of Iowa, Illinois, Wisconsin, and Missouri combined, Ukraine boasts 34 million hectares of arable land, roughly a third of the amount farmed in all the European Union. (It lost almost 2 million hectares of agricultural land when Russia seized Crimea in 2014.) Buried beneath nearly 70 percent of that, however, lies Ukraine's national treasure: some of the richest soil on the planet.

The thick, dark *chernozem* (Russian for "black earth") is 6 meters deep in places and rolls out in a broad swath from western Ukraine's Carpathian Mountains to the southern Urals of Kazakhstan. Similar to the soils of the US Midwest, the black-earth belt was laid down as loess, the wind-borne silt ground by Pleistocene glaciers that gave rise to Earth's great grasslands—the Great Plains in North America, the pampas of Argentina, and the steppes of eastern Europe. Millennia after millennia of deep-rooted perennial grasses, scorched by the occasional wildfire, created a fertile soil full of organic matter and perfect for growing crops, especially cold-loving grasses like wheat. With a climate and topography not too different from those of Kansas, Ukraine covers one of the finest wheat-growing regions of the world.

Unlike Kansas or Iowa, which achieve some of the world's highest yields, Ukraine's agricultural production plummeted after the fall of the Soviet Union in 1991. The country suffered "one of the most severe and prolonged economic declines of any economy in the Former Soviet Union or Eastern Europe," according to the World Bank. It has never recovered. Today Ukraine's farmers typically harvest about half the amount of corn, wheat, barley, and rye that they did in the mid-1980s. Ukraine is more famous for the 1986 Chernobyl nuclear disaster than for silos full of grain. Though 150,000 square kilometers of northeastern Ukraine remains contaminated with radioactive strontium-90 and cesium-137, prevailing northerly winds carried most of the Chernobyl fallout to farmlands in neighboring Russia and Belarus. Still, 8 percent of Ukraine's agricultural lands remain too radioactive to farm.

Even so, Ukraine has enormous potential for helping to solve the global food crisis, if it can transform itself from basket case back to breadbasket. Agricultural experts at the FAO, World Bank, and OECD have long maintained that if the former Soviet states of Russia, Ukraine, and Kazakhstan could achieve yields close to the old Soviet levels of production—to close the "yield gap" between their harvests and those of the United States, Europe, and China—the troika of potential wheat powerhouses could significantly ease the global grain crisis.

Even with its poor wheat yields, Ukraine has become the fifth-largest exporter of wheat in the world, selling about 5–10 million tons each year, mostly to North Africa and the Middle East. According to agriculture experts at the World Bank, Ukrainian soils could easily double their average yields, from 2.8 tons per hectare to 6, in good years. At that level, they could produce 42 million tons. Take 12 million off for domestic consumption, and Ukraine could still export 30 million tons—20 percent of the average annual world trade in the grain. Moreover, while arable land is shrinking in other breadbaskets, less than half of Eurasia's fertile *chernozem* is currently planted in crops. Ukraine alone has 3 million hectares of fallow farmland. This is one of the biggest wastes of agricultural potential on the planet.

UNFORTUNATELY, UKRAINE'S FECKLESS politicians have failed to shake off Soviet torpor. Instead of alleviating the food price crisis of 2008, Ukrainian leaders exacerbated it. A severe drought in 2007, combined with rising food prices and political turmoil at home, led the government to place export restrictions on grain and sunflower seed oil. In 2008, however, Ukrainian farmers reaped a bumper crop—more than 53 million tons of grain—for the first time ever approaching yields of the Soviet era. Yet because of the export restrictions and lack of grain elevators, they were forced to dump $100 million of spoiled wheat into the Black Sea.

That incredible agricultural potential, together with the record grain prices during the food crisis of 2008, is what lured Justin Bruch to Ukraine, along with a flurry of international investors hoping to snap up cheap agricultural land. But here the global landgrab took on a typical Ukrainian twist. After the fall of the Soviet Union, the new nation began a lengthy process of land privatization, giving land to former farmworkers. As a result, millions of rural villagers now own tens of millions of rectangular plots that average about 2 hectares each. But under a landmark reform law passed in the early 2000s, they were barred from selling it for 10 years; they could only farm or lease it. What has become known as "the moratorium" on land sales has been extended several times and is now set to expire in January 2016.

One provision of the law gave lessees the right of first refusal to buy the land. And since Ukrainian farmland is expected to sell up to 20 times *cheaper* than farmland in nearby Europe, both international investors and Ukrainian land companies began leasing huge tracts from the impoverished Ukrainian peasants. Meanwhile, EU biofuel mandates drove up the demand and prices for oil seeds like rape and sunflower, both of which thrive in Ukraine. Soon everyone from traditional grain giants like Cargill, Bunge, and Glencore to investment banks like Morgan Stanley and Russia's Renaissance Group were pouring money into Ukraine, part of an estimated $10 billion in direct foreign investment in the country's agricultural sector since 2005. Many had no idea how to farm. But the Ukrainian land rush was on.

"They thought that this was the next dot-com thing," said Bruch, who was hired by Morgan Stanley to manage a 3,000-hectare farm in southern Ukraine, one of 11 the company had acquired in 2008. "People thought this deal was so good they were fighting to get in on it. A guy high up in Morgan Stanley told me, we don't need to make money farming. We just need not to lose too much. It's a land play. We're going to be able to buy that land in a few years at $600 an acre and sell it for four times that. Plus, they thought any idiot can grow crops. They were going to manage the whole thing from their desk in New York."

It didn't quite work out that way. Instead, Morgan Stanley lost an estimated $30 million on the deal before bailing out in 2009 after the financial crisis. A British company named Landkom, which was one of the first foreign investment groups to ignite the land rush in 2007, lost more than $111 million before it was bought out by Bruch's current employer, Swedish investment group Alpcot Agro (now known as Agrokultura). The combined company now controls more than 68,000 hectares of farmland in Ukraine and 158,000 in Russia.

ON A SUNNY September morning I climbed into Bruch's shiny black Toyota Land Cruiser, and soon we were wending our way through the narrow cobblestoned streets, dodging pedestrians and old Soviet trolleys packed with morning commuters. Bruch apologized for the car. He'd had his own Dodge 4×4 crew-cab pickup shipped over when he worked for Morgan Stanley. But it was such an oddity on the road and such an obviously American vehicle that it had become a police magnet. The Ukrainian cops would stop him just to see what the inside of the big Dodge looked like. And they always demanded bribes—usually 100 hryvnia, Ukraine's unpronounceable currency, or about $12. The money didn't bother him, he says. Graft is essentially part of the police salary here. But the wasted time did.

Bruch now manages 60,000 hectares of land, or roughly 150,000 acres. That's an area about the size of Chicago, but spread out in clumps within a 150-kilometer circle of Lviv. Even with some 800 employees, trying to grow three winter crops (wheat, rape, and barley) and three spring crops (sunflower, soybeans, and buckwheat) on that much land is a constant race. Bruch typically spends 14 hours a day driving from farm to farm. If he's not talking on his iPhone, its steam-whistle ring tone is soon blaring. At times it's hard for him to finish a sentence before the whistle blows. When Alpcot took over Landkom, the Land Cruiser came with the deal. Since he started driving it, he's been pulled over only once (because the car was so splattered with farm mud that the police thought it had been stolen), which has been a huge time-saver and a small window into the Ukrainian mind. "In Ukraine, the bigger and blacker the vehicle you drive, the more important you are," said Bruch with a laugh.

The road grew smaller and more potholed as we left Lviv behind. The land opened and spread out in gently rolling hills bordered by dense thickets of trees in the bottoms. Bruch said the country-side reminded him a bit of Wisconsin, with its patchwork of small pastures dotted with cows. In order to bring the green revolution back to Ukraine, though, groups like Alpcot Agro have to cobble enough land together to create economies of scale—essentially recollectivizing the old Soviet-style farms. Ukraine's fiercely inde-pendent peasants were among the last collectivized by Josef Stalin, and many still accuse the ruthless Soviet dictator of manufactur-ing the horrific famine of 1932 that killed millions.

These days, if Bruch wants to lease 100 hectares he goes to the local village council and at the next meeting they vote on the pro-posal. The government sets the minimum rent of 3 percent of the land's assessed value, although like everything else in the country, that's negotiable; Bruch's rents average about $80–$100 per hect-are. If the village council agrees to lease the land, Bruch will add 50 new landlords to the tens of thousands he already has. Most demand to be paid in gunnysacks of grain, since severe bouts of hyperinflation have left them leery of the hryvnia. "In these little villages they don't trust government and they don't trust the cur-rency," Bruch said. "In grain they trust."

We drove through small towns and picturesque villages, most with statues of a glowering Taras Shevchenko, the nineteenth-century poet, artist, and serf-born revolutionary now deemed the father of Ukraine. The statues peered out on villages not terri-bly different from those in Shevchenko's bucolic paintings. Neatly kept brick and stucco houses lined the roads, their yards dappled with flowers, apple trees, and chickens. Ancient babushkas in long dresses and bright head scarves urged bony cattle down the roads with a switch, while young men drove carts pulled by skinny horses, a cell phone their only nod to modernity. It was harvest time in Ukraine, and we passed many families working rich, black land dotted by small shocks of corn drying for fodder—a method now seen in the United States only on Amish farms or in pumpkin displays. Potbellied Papa, sweating in black pants and no shirt, typically dug the potatoes with a short-handled potato fork, while Grandma, Ma, and children of all ages placed the tubers in white gunnysacks, either to store for the winter or to sell at the local market.

On the surface it all seemed quaint and almost Bavarian, until I noticed the colorfully painted little well houses with the pointed roofs in the yards. In 2011 only a third of rural households in Ukraine had running water or sewer, and only 6 percent had a hot-water supply. That means hauling water, breaking ice, and using outhouses during the bitter Ukrainian winter, when temperatures can fall to 30°C below freezing. With 36 percent of rural families below the poverty line, Ukrainian villages are in population free fall, increasingly occupied by aging peasants whose children have left to seek better lives in the cities or abroad.

The little garden plots—and the Ukrainian peasants' agricultural skills—have been critical to their survival. After the Soviet Union fell, the entire nation was impoverished overnight. Food prices increased 17-fold between December 1991 and December 1992, and they continued to spiral upward well into the 1990s. Per capita consumption of meat and dairy products fell by half during the decade. By 2001 the household plots were producing 77 percent of the meat, 73 percent of the milk, 87 percent of the vegetables, and 98 percent of all the potatoes consumed in the entire country. That year 83 percent of the population was getting by on what was deemed the bare minimum of about 311 hryvnia ($58) each month. The situation has improved only slightly since.

Even the larger Ukrainian-owned farms are struggling, with poor inputs, old machinery, less-than-stellar seeds, and a 1970s-era farming mentality. "They are doing almost 3 tons per hectare wheat and 6 tons of corn on average," said Bruch, as he rolled to a stop outside Alpcot's big farm base in the village of Rohatyn. "We're doing 4.5 tons per hectare wheat and 8.5 tons corn."

The base was a long brick warehouse beside an abandoned railroad spur—the former fertilizer depot for the surrounding collective farms. Parked in a long line outside stood the tools of Big Ag: two 12-meter-wide AgCo air seeders shipped over from North Dakota, along with a 16-row Kinze corn planter from Iowa—machines so wide that their articulated arms have to fold up on themselves so they can travel down the road. Muscular green and red four-wheel-drive tractors from John Deere and Case IH stood beside giant John Deere harvesters and a flotilla of trucks, grain carts, sprayers, water tanks, and other tools of a corporate farmer's world.

Inside the building Bruch introduced me to the farm's management team: Vladimir Bubnov, a young Ukrainian with a blond crew cut and chiseled Teutonic face; and his agronomist, Andrey Portrylo, who sported a big smile and the requisite Ukrainian beer belly. Jimmy Zimmerhanzel, a grain elevator expert whom Bruch had brought over from Taylor, Texas, was also there that morning helping out. Together they explained the first problem of the day. The warehouse was filled with dunes of winter wheat that had been saved for seed from last year's harvest. The wheat needed to be treated with fungicide, bagged, and shipped out to the farms for planting to make room for the corn harvest that would be starting any day. The small Russian seed treatment machine, however, wasn't working. A half-dozen men stood watching while two others banged on its clogged grain auger with hammers and wrenches, making a god-awful racket. Zimmerhanzel was on it and soon, through an interpreter and hand gestures, had the men opening the auger and turning it by hand to clear the blockage.

"My job is to make things work better," Zimmerhanzel said, his gray T-shirt already soaked with sweat at nine o'clock in the morning. "A lot of these guys, they're going to get their paycheck at the end of the month. They just don't care." When he first came to Ukraine, the company's two grain elevators, with a total capacity of 50,000 metric tons, were each processing about 400 tons a day. One of the elevator managers told Zimmerhanzel that was all they could do—that it was impossible to do anymore. Zimmerhanzel explained to him that if he lined up the trucks in a certain way and added an extra shift, he could do a lot more. The manager refused. So Bruch fired him and implemented Zimmerhanzel's suggestions. The two elevators now process 1,800 tons a day, and Bruch is negotiating with his bosses to build a third.

"This is one of the hardest cultures I've ever seen to change," Bruch said. "It's just extremely difficult. You explain something to people here, and they'll fight you over it. It will be fabulously successful, and then the next year you have to fight them all over again."

We headed over to a nearby wheat field where things were going smoothly. Two huge tractors were pulling big German Horsch planters, drilling wheat into a field of rape stubble that rolled to the horizon. Bruch's company leases as much contiguous land as possible to allow him to use bigger tractors, planters, sprayers, and harvesters that involve less time, less labor, and lower costs of production per hectare.

Bubnov motioned the driver of the big Case tractor to stop and take a break, and he and I climbed into the air-conditioned cab, closed the airtight door, and rolled across the field with eastern European techno-pop thrumming over the muffled sound of the big diesel engine. Bubnov, at 30, is part of the new generation of Ukrainians who are trying to westernize the country. Though he grew up south of Kiev and trained as a veterinarian, there was little money or opportunity in Ukraine when he graduated. So he went to Europe, where he worked on vegetable farms and hog operations in England and Sweden. He had worked for Bruch's consulting firm for the last five years and had become Bruch's right-hand man. With his John Deere cap, Oakley Blades, and Leatherman on his belt, he could have been any young farmer from Nebraska.

Like most of Alpcot Agro's equipment, the tractor brimmed with the latest technology, including GPS autosteering and a laser mounted on the planter to ensure that the rows were perfectly straight. The tractor cockpit had so many gauges and buttons that it looked like an F-14, replete with a computer screen monitoring every seed hose. The big Horsch planter hummed as it laid out a wide swath of shallow slits through the stubble of the brown field behind us, each one filled with seeds at the precise depth and spacing for maximum emergence and minimum disturbance of the soil.

"I really like no-till," says Bubnov in perfect English with a slight Ukrainian accent. "It's my idea of proper use of the land. It saves everything—fuel, seeds, soil. Maybe we get slightly lower yields, but we use less inputs, which saves money. But in this area people like to do tillage, and it's not easy to break them of that habit." The problem was particularly acute when Bubnov and Bruch were farming for Morgan Stanley in southern Ukraine. That region is far drier than the area around Lviv, receiving only about 12 inches of rain each year. To Bruch it made absolutely no sense to break land in dry country, since plowing releases any soil moisture, but the farmers there were adamant that no-till wouldn't work, even after he and Bubnov tried it and got the best yields of any farm in the region.

Corruption is another major issue. Bruch employs a small army of accountants and security personnel to make sure nothing walks away, even zip-tying fuel caps on the vehicles to make sure no one steals a few liters of diesel. Bruch's team has a running joke about a former agronomist at one of the farms whom they nicknamed

"Big Fat Petrovich." Once when they were walking through a wheat field that was almost ready for harvest, Bruch and Petrovich were discussing what they thought it would yield. Bruch thought it looked pretty good, perhaps close to 4 tons per hectare. Petrovich said, "No. Yield will be 2.7 tons per hectare." And sure enough, the wheat yield on the farm came out to exactly 2.7 tons per hectare.

"It was uncanny," Bruch said. "I never knew an agronomist who could call a yield on the nose. Of course, I found out later what he meant was that everything over 2.7 tons was his. There's an old saying here that goes something like, 'You steal a bag of wheat, you go to jail. You steal a truckload of wheat, you go on vacation.' It's not that they're bad people, just the Soviet system they were raised under."

In fact, the work ethic is so bad in Ukraine that it even factored into a major report in 2001 from the OECD and the World Bank, groups that usually show more tact:

> The former organization of labour, with an abundant labour force and limited individual responsibility, has left a legacy of poor motivation and appalling work practices.

The report also excoriated management on the large Ukrainian farms. But management is poor at all levels of government. In 2013, Ukraine ranked 144th out of 177 countries and territories on Transparency International's Corruption Perceptions Index, on par with Papua New Guinea and the Central African Republic. Anger over persistent government corruption was one of the factors that led to the 2014 ouster of President Viktor Yanukovych, who fled to Russia after opposition groups sacked police stations and a military base in Lviv, Bruch's adopted home.

After making a few passes across the field, Bubnov turned the tractor over to the driver and we headed into Rohatyn for lunch at a local café with the rest of Bruch's team. Portrylo ordered a traditional meal for the table, as well as a round of *horilka*, the legendary Ukrainian vodka. "For guests," he said, pouring three shot glasses full. "It is tradition."

The vodka was smooth and soft, lighting a warming fire from my toes to the thinning hair on the top of my head. Vodka is a Cossack drink, invented and some would say perfected in Ukraine. It arrived on the table with a plate of *salo*—what we would call fatback in the South—cubes of pork fat topped with slivers of garlic. Rumor has it that Ukrainians think eating *salo* is better than sex. True or not, consuming *horilka* and *salo* together in Ukraine is akin to taking the sacrament.

The *salo* was followed by a creamy pink borscht full of shaved beets and dill, perhaps the best soup I have ever tasted, and small but filling potato pancakes called *dramaki*. As we ate, I asked Bubnov and Portrylo what it had been like in Ukraine after the fall of the Soviet Union. Portrylo, whose English was sparse, just shook his head and said, "Very bad." Bubnov provided more details:

"When the Soviet Union fell in 1990–91, there was high inflation," he said. "Today bread was 8 hryvnia, tomorrow it was 20. For 8–10 years nearly all the big collective farms went fallow. During that time they privatized the land. Collective farms had equipment, but it was not for the people's use. When the collectives went broke, banks took the equipment to sell. So the people couldn't do anything on it. They just had the land. In the south the USSR built a large irrigation system, with big canals like California. Once the Soviet Union collapsed, the irrigation channels deteriorated. Villagers stole the pipes."

After lunch Bubnov jumped behind the wheel, and we flew down the potholed rural blacktops to the next farm as if we'd just entered the Paris–Dakar Rally with horse carts serving as living obstacle cones. On better sections of road we passed slower vehicles six at a time at 140 kilometers per hour (almost 90 miles per hour). We crossed a bridge fording the wide Dniester River, one of Ukraine's great waterways, as well as a behemoth power plant with three giant smokestacks pouring black soot into the air. The plant, Portrylo said, had been privatized and sold to one of Ukraine's industrial oligarchs, who then sold all the power to Poland.

The next farm had a new problem. A recent windstorm had damaged 1,000 hectares of sunflowers just before harvest. I had envisioned a sea of giant black-eyed Susans stretching for the sun. Instead, the storm had left the field a tangled sea of brown, with seed heads the size of pie plates all pointing to the ground. The harvester was having trouble picking them up with its huge header—a rolling, tined reel that gathered them into the cutter bar. The farm's mechanic had fitted the header with long pointed metal pans to catch the seeds that broke off early, but they weren't working very well in the tangled crop.

The farm manager here was a young Ukrainian businesswoman named Lisa Potter. She had no agricultural experience; she had been a human resources executive in England (where she had acquired an English husband and surname) before coming home. But she knew how to manage people and was trustworthy, perhaps the most important requirement for foreign companies operating in Ukraine. Her solution was to hire a dozen women from the village to cut the sunflowers by hand, paying them 100 hryvnia a day to stack the stalks in carts that would be fed into the combine. The sight of the middle-aged babushkas in their dresses and scarves going up against a German Claas Lexion 600 capable of harvesting 60 tons of grain per hour was almost painful to watch, and it seemed somehow symbolic of Ukraine's agricultural metamorphosis. Bruch just shook his head, but Potter explained it was good PR for them in the village. Bruch suggested they take off the pans and try a corn header—which looks like a fat, multipronged metal fork—on the sunflowers the next day. It worked well, and the village women were let go.

The labor situation is tense. In the Soviet days, everyone in the village had a job at the collective, whether they were needed or not, and the farm provided social services like health care. Nearly every Western agricultural technology or method developed since the green revolution is designed to save labor for the farmer—and thus is a potential job destroyer for these villages. Bruch's no-till battle in southern Ukraine was a classic example. No-till planting typically requires far fewer passes over the field and thus far fewer tractor drivers. Since most villagers have little more than elementary school educations, it's difficult for them to find jobs in the cities. Simply finding Ukrainian workers capable of driving Bruch's modern tractors, sprayers, and harvesters with their raft of computers and electronics is a considerable challenge.

The huge, high-tech equipment, along with heavy use of chemicals, is the only thing that makes farming such a large swath of land remotely feasible, and Bruch tries to keep the big machines running around the clock during the busy seasons in spring and fall. But it doesn't always work that way. On the way to inspect a former field of rape that was being sprayed with herbicides to prepare it for planting wheat, we passed two of the company's huge tractors parked beside one of the tractor drivers' houses with their gangly cultivators neatly folded behind them. Bruch was immediately on the phone to the farm's manager.

"We don't have time to have equipment sitting around. We're 10 days late already!"

"They have no fuel." Came the reply.

"Well, request an invoice and get some fuel! It makes no goddamn sense for those machines to be sitting still. They need to be rolling now!"

He relaxed a bit when we reached the field and saw a red-and-white sprayer flying over the land like a giant jacked-up dune buggy with wings. It was an American-built Miller "Nitro," the Cadillac of sprayers, Bruch explained, with a 120-foot boom, 4,000-liter tank, GPS autosteering, and "autoswath," a device that utilizes GPS to snap on and snap off nozzles at the end of the row or as the machine rolls around obstacles in the field to prevent overspraying. With the GPS, it can even steer itself around the field; all the operator has to do is push on the joystick throttle. The Cadillac of sprayers costs as much as 8 new Cadillacs, or roughly $400,000. Bruch's company has two of them in its fleet of 20 sprayers.

"The most important thing you can have over here is a good sprayer," Bruch says. "It's so easy to get behind when you are farming this much land. That Nitro will run 20 miles per hour, which amounts to 72 hectares an hour. My target is to spray 500 hectares a day."

But again, finding skilled operators that can maximize the machine's time is hard, as is repairing them when something inevitably breaks. "The Russian guys who farm here run old Belarus and Russian equipment, and it has its advantages," said Bruch. "Everyone here can drive it and everyone here can fix it. You can buy a cylinder at the local gas station. If the hydro goes out on the Miller, I have to get a part from the States and that could take weeks."

The one modern agricultural technology Bruch doesn't use is GMO seeds, which have been banned by the Ukrainian government, although Bruch believes most of the soybeans grown in Ukraine come from "Roundup Ready" varieties that were smuggled into the country during the 1990s. Several tests of Ukrainian soybeans and poultry products in the European Union have confirmed the existence of the trait. Bruch avoids them because, as a foreigner, he doesn't want to run the risk of being caught.

Not all the crops that Bruch grows are chemical-intensive. He plants a lot of spring buckwheat (*Fagopyrum sagittatum* Gilib), which is not wheat at all, but rather a close cousin of rhubarb. It takes little fertilizer, grows so fast it smothers most weeds, makes a decent manure crop when plowed under, and has a higher nutritional content than any cereal grain. It's even gluten-free. Ukrainians love it on the table and love to see its auburn seed heads growing in their fields. They call it *gretchka* and believe it's one of the healthiest things you can eat. I had a delicious bowl of *gretchka* mixed with yogurt every morning I was there. It's not as profitable as sunflower or rapeseed, says Bruch, but during the food price crisis it rose to $1,000 per ton. He was producing 1 ton per hectare, but since it cost him only $300–$400 to grow, he was making good money on it. Plus, it's a short-season spring and summer crop that can be harvested in 75 days, making it a potential double crop with winter wheat. With better shorter-season seeds, farms could potentially grow a crop of wheat for export and a crop of buckwheat for local consumption on the same field in the same year.

In fact, Ukraine has attracted several large investors that hope to grow organic produce and cereals for the European market, including the group founded by the Beatles' spiritual adviser, the late Maharishi Mahesh Yogi, who transformed his teachings of transcendental meditation to a global real estate empire that now includes 50,000 hectares of Ukrainian farmland. After several years of deliberation, in August 2012 the Ukraine parliament finally passed an organic-agriculture standards act equivalent to the European Union's. The country now has an estimated 270,000 hectares of certified organic farmland.

Though most of the organic crops so far have been grains and oilseeds, Swiss and German development agencies have been working to help small family farms tap into Europe's burgeoning organic produce market. Many of Ukraine's vegetables and fruits are already grown with few chemical inputs, simply because most Ukrainian small farmers can't afford them. The premium prices brought by organic produce could help lift millions of Ukrainian farmers out of poverty. The potential for Ukraine to be the organic farm of Europe may even rival its potential to become a major grain exporter to the world.

The challenge for Ukraine is to avoid the pitfalls of the green revolution as it ramps up its agriculture. Much of Ukraine's farmland already suffers from extensive erosion caused by excessive tillage on sloping fields. Nutrient runoff from poor fertilizer and manuring practices pollutes every water body and contributes to algal blooms in the Black Sea. According to a World Bank report in 2007, a quarter of the farmland is contaminated with pesticide residues from the Soviet days, with stockpiles of old, obsolete chemicals creating a potential health hazard. Ironically, during the Soviet era Ukraine was a leader in developing predatory insects and other biological controls for crop pests. Since the amount of agricultural chemicals used is an order of magnitude lower than during Soviet times, and much of the land remains fallow, Ukraine has an enormous opportunity to farm better the second time around. By adopting a nationwide program of conservation tillage, nutrient and pest management programs, and established buffer zones and wetlands to keep fertilizers out of streams, Ukraine could create an agrarian model for the rest of the developing world to follow. Companies like Agrokultura could serve as training grounds for the next generation of Ukrainian farmers.

UNFORTUNATELY, SHEVCHENKO'S BELOVED homeland is headed in a far darker direction at the moment. Pro-Russian separatists have mined the roadsides in the areas they control in the east, preventing farmers from harvesting their fields; and tensions with Russia over its invasion of Crimea could throttle agricultural exports from Ukrainian ports. Hostilities with Russia aside, the biggest stumbling block to Ukraine's agricultural revolution remains the lack of a free and fair market in agricultural land. Ukrainian leaders initially had good reason to fear that oligarchs and foreign corporations—like Morgan Stanley—would attempt to buy huge swaths of the countryside in speculative land deals, perhaps with coercive sales at rock-bottom prices. But without proper title or the ability to buy and sell, no farmer—large or small—will invest in the improvements needed to make the farms more productive and able to reach their potential.

One of the oldest maxims in agriculture, if not all real estate, is that no tenant will care for his property as lovingly as an owner will. The European Union and World Bank are trying to help Ukraine make the transition when the current moratorium on land sales expires in January 2016. But what Ukraine's fractious parliament will do at that point is anyone's guess. On paper, the country has everything it needs to produce a river of grain flowing to the hungry parts of the globe. The only things standing in its way are civil war, corruption, and a Communist hangover that seems to have no cure.

After the last field was inspected, and the last problem of the day resolved, Bruch, Bubnov, Portrylo, and I climbed back into the shiny black Land Cruiser for the two-hour drive back to Lviv in the gathering dark. It had been a long day, and Bubnov drove hard and fast through the night. We talked about the future of Ukraine, if it will ever fulfill its agricultural destiny, and what it might look like if it does. Bubnov, surprisingly, didn't think corporate farms would take over. They are too inefficient, he said. He thinks that one day Ukraine will be dominated by medium-sized family farms like those in the United States and Europe. Bruch agreed.

"If the Ukrainian government can get this right, they could be a First World country overnight. Better than Brazil," Bruch said. "If you could get free title and deed to the land, and debt financing, this country would be incredible. But if they screw it up, that's it. They're screwed for the next century."

A Bangladeshi farmer waters his crop by hand during a severe drought in 1972.

CHAPTER 10

The Blooming Desert

The beast of the field shall honour me, the dragons and the owls: because I give waters in the wilderness, rivers in the desert, to give drink to my people, my chosen.—*Isaiah 43:20*

Several years ago, while reporting a story on California's Salton Sea, I spent a morning riding around with Lloyd Allen, a 75-year-old farmer whose family had moved to the Golden State from Arkansas during the Dust Bowl because, he said, "we were hungry." At one point we stepped out of his truck and climbed a small brown hill a few miles from his home in Calipatria, California. In front of us, brown desert scrub stretched out to the Chocolate Mountains in the east, where navy jets routinely drop thousand-pound bombs. The battered hills of the gunnery range flowed south to the giant Algodones Dunes—a 40-mile-long sandpit where each weekend thousands of off-roaders in dune buggies, dirt bikes, and monster trucks do their level best to kill themselves with their machines.

It is a landscape so bleak and lifeless that early settlers dubbed it "Valley of the Dead." A place so arid and barren that NASA trained Apollo astronauts here for their missions to the moon.

We then reached a small ditch, turned to face west, and the contrast couldn't have been more stark. Lush, green rectangles marched in orderly rows across a table-flat valley bisected by dirt roads that met at perfect right angles. Spinach, potatoes, broccoli, onions, sweet corn, cauliflower, lettuce, asparagus, alfalfa, wheat, cotton, and a hundred other crops carpeted nearly a half-million acres. The Valley of the Dead had been transformed into the Imperial Valley, a billion-dollar agricultural economy now known as the "winter salad bowl" of the United States. Two-thirds of the nation's winter vegetables are grown here.

All these crops thrive in a place that is technically the Sonoran Desert, receiving on average just 2.3 inches of rain each year. The secret ingredient, of course, is irrigation water—about 3 million acre-feet to be precise—a sliver of which was flowing through the concrete ditch at our feet called the East Highline Canal. The clear-green life-giving liquid bubbling through the trough had traveled a thousand miles to get here. It had fallen as snow on the peaks of the central Rocky Mountains, where it eventually melted, tumbling down rivulet, stream, and tributary to join the Colorado River. It had joined the brawny rapids in the Grand Canyon, drained through the slack water of Lake Mead, and churned through the turbines at the Hoover Dam lighting Las Vegas. From there it flowed downriver to Yuma, Arizona, where the Imperial Dam diverts 20 percent of the Colorado's flow into the All-American Canal. The 80-mile lifeline of the valley is the far less famous leg of the 1928 Boulder Canyon Project that built the Hoover Dam, but it has become infamous in its own right. It is the largest irrigation channel in the world, 700 feet wide and 50 feet deep in places, running through desert and dune near the Mexican border. It is also among the most deadly. More than 500 people have drowned trying to swim across it to enter the United States.

The All-American is part of 3,000 miles of canals and drainage ditches that water nine cities and those half-million acres within the Imperial Irrigation District (IID), the largest irrigation district in the United States. My companion, Lloyd Allen, farmed 4,000 of those acres. He had served on the IID's board of directors for nearly 20 years, and when I met him he was also the president of California's Colorado River Board, the group that jealously guards the state's century-old water rights to the river. He was known in water circles as a plain-spoken, hard-nosed negotiator. He looked over the green fields covering the desert valley below us like Moses taking stock of the promised land. He knew it would all be a mirage without the water.

"This is one of the few places on the planet where you can call God on the phone and say I want 12 feet of water and he gives it to you," Allen told me as we stood beside the canal that day. "And God is the IID."

Agriculture is a thirsty business. The "foot" of water that Allen referenced is an acre-foot—the amount needed to flood an acre of land (43,560 square feet, about half of a hectare) to a depth of 1 foot. That amounts to nearly 326,000 gallons, enough to supply one typical US suburban household for a year, or two households in water-thrifty California. The water used each year by 450 Imperial Valley farmers could supply half the homes in California.

In fact, of all the freshwater available for human use around the world, agriculture sucks down nearly 70 percent, or about 2.7

million cubic meters (enough to fill a canal 100 meters wide, 10 meters deep, circling the globe 180 times). In the United States, agriculture uses closer to 80 percent of the water. We've stored so much water behind dams in lakes and reservoirs that the collective weight has created a measurable wobble in the spin of the planet.

The reason should be obvious to anyone with a garden hose and a tomato plant on a hot summer day. Humans are born irrigators. It's no coincidence that the first city-states of Mesopotamia, the birthplace of agriculture and all civilization, arose between two great rivers, or that their foundations were laid by an invention just as revolutionary as the wheel: the lowly irrigation ditch. Imagine the thrill of the first person who cut the bank of the Tigris during a drought and saw the life-giving water flow into his or her fields. In nearly every place with fertile soil, little rain, and a handy river—the Nile, the Yellow, the Indus, the Mekong, the Columbia, the Yangtze, the Colorado—farmers, settlers, and politicians have built dams, diversions, ditches, and canals. Aside from beating back drought—farmers' perennial foe—dams bring other politically popular benefits, including construction jobs, flood control, water for growing cities, and cheap electricity, not to mention a seemingly inexhaustible supply of kickbacks and bribes.

The environmental cost of turning free-flowing rivers into slack-water ponds, however, has often been horrendous. Up to 15 million salmon once returned each year to the Columbia River basin, an abundant renewable food resource that fed the Native Americans of the Pacific Northwest for millennia. The first irrigation diversion went up in the region in 1843. There are now more than 450 dams on the Columbia and its tributaries, providing flood control, irrigation, and more than half of the electric power for the entire Pacific Northwest. Though they weren't the only factor in the fall of the salmon—overfishing, logging, and agricultural runoff all contributed—the dams blocked off more than half of the salmon's former spawning habitat. Today, despite spending nearly a half-billion dollars annually to aid their recovery, only 100,000–300,000 wild Columbia basin salmon remain—less than 2 percent of their historic population. Nearly all of the stocks (27 at last count) are listed as threatened or endangered under the federal Endangered Species Act. In fact, all over the globe freshwater animals are going extinct five times faster than those that live on land or in the sea, primarily through habitat loss caused by our insatiable demand for water.

The dam builders believe the loss of millions of salmon was a small price to pay. In a 1994 interview, 84-year-old Floyd Dominy, the cigar-chomping former director of the Bureau of Reclamation who presided over the last spate of dam building in the United States during the 1950s and 1960s, unabashedly boasted of his achievements: "I'm sure people can survive without salmon," Dominy said. "But I don't think they can survive without beans and potatoes and lettuce. . . . I think the [dams] were worth it. I think there's substitutions for salmon. You can eat cake."

In terms of global food production, the impact of all that irrigation has been nothing short of miraculous. In the developing world, crop yields are two to three times higher under irrigation than in rain-fed fields, and cropping intensity—the number of harvests in a given year—is more than 50 percent higher. In Vietnam and Indonesia, farmers with irrigation can grow as many as seven rice crops in two years—compared to two to four crops for rain-fed rice. Even though only 20 percent of the world's farmland is under irrigation, those 300 million hectares put 40 percent of the food on the world's tables each year.

The bulk of that irrigation was built after World War II, with investments in dams, canals, and other infrastructure peaking in 1970, the year Norman Borlaug won the Nobel Peace Prize. The timing was no coincidence. While most people think of the green revolution as a package of semidwarf wheat and rice and copious amounts of fertilizers and pesticides, none of that would have come to much without timely irrigation to maximize the uptake of nitrogen and empower the new seeds to do their thing. Since 1950, when the green revolution was just beginning to bloom in the irrigated fields of Mexico, the amount of irrigated farmland has doubled, while our water withdrawals have tripled. When agricultural experts say we need to "intensify" production, what they often mean is to expand irrigation.

Just as humans sweat more on hot summer days, plants will need more water as climate change raises temperatures. The process of water use in plants, called "evapotranspiration," is perhaps the most important biological process on the planet after oxygen-producing photosynthesis. Not only is it necessary for the production of all plant biomass, but evapotranspiration is a critical component of Earth's great hydrologic cycle and the climate itself. As the sun warms the leaves, driving photosynthesis, the tiny pores on the leaf's surface known as the stomata open. The surface tension of water molecules, combined with capillary action through the hollow xylem tissue of vascular plants, draws water from the roots to the crown like a giant straw sucked by the sun. In the Imperial Valley, part of the broiling Sonoran Desert, the evapotranspiration rate for thirsty crops like alfalfa is 82 inches a year, more than 30 times the average rainfall in the valley.

A warmer climate also means more water vapor in the skies, which translates into harder rainfall and less snow and ice collecting on mountain peaks. All that picturesque frozen water acts as a gigantic vertical reservoir. Throughout most of agricultural history, the snowpack and glaciers of the world's great mountain ranges have melted slowly throughout the summer—conveniently, the part of the growing season when farmers downstream need water the most. Meltwater from the Himalaya, the Tien Shan, and the Rockies, among other ranges, currently supplies 40 percent of the world's irrigated farmland. A future with less snow and ice means more runoff in winter and spring, when crops are small or not yet planted, and less runoff in summer, when they need it most.

Already, peak runoff is occurring earlier and reducing summer flows in the Columbia River basin. Even rainy Washington State's Yakima Valley, the leading US producer of apples, sweet cherries, and hops, that critical ingredient of beer, is headed for drier times. Though the valley's Cascade Mountain headwaters receive a whopping 80–140 inches (203–356 centimeters) of precipitation every year, earlier snowmelt will likely cause increasingly frequent water shortages among valley farmers with junior water rights. But farmers with senior water rights will likely have their allotment cut in dry years as well—something that hasn't occurred in more than a century of irrigated farming in the valley.

According to the IPCC, the impact of climate change on the world's freshwater supplies will likely vary widely from region to region, with arid and semi-arid regions getting drier and other areas experiencing heavier, more variable rainfall and more frequent flooding. They predict the negative impacts will outweigh any benefits.

Less snowfall and depleted aquifers present the farmers of the future with a huge challenge. A recent study by the FAO estimated that our increasing population and incomes around the world will require an 11 percent increase in irrigation withdrawal by mid-century. If you add the hotter temperatures of climate change to the mix, the problem starts to look like the curse of the Danaides, the 49 brides of Greek mythology who slew their grooms and spent eternity hauling water in a fruitless attempt to fill a basin full of holes. The math is simple: a 1°C rise in temperature results in roughly a 3–4 percent increase in water demand. The FAO predicts that climate change will cause global irrigation demand to increase by as much as 20 percent. A more detailed study by the

USDA in 2008 projected a 35–64 percent increase in demand in the United States by just 2030. In perhaps the most disturbing estimate, after spending five years putting together a *Comprehensive Assessment of Water Management in Agriculture*, 700 of the world's top water experts concluded that, without significant gains in productivity in the world's rain-fed fields, water demand by agriculture will nearly double by 2050—a quixotic vision, since the water for such an expansion just doesn't exist.

Not that governments haven't tried. The Chinese completed their colossal and controversial Three Gorges Dam in 2006. Critics contend that the sheer weight of the water in the world's largest hydroelectric project has led to landslides and earthquakes—like the devastating 2008 Sichuan quake that killed some 87,000 Chinese. The graft involved in such megaprojects is staggering. The cost of the Yacyretá Dam on the Paraná River between Paraguay and Argentina rose from an initial estimate of $1.6 billion to $8 billion, earning it the moniker "monument to corruption." Nor is such chicanery limited to the developing world. In 2011, two program managers for the US Army Corps of Engineers, the primary water project agency in the United States, were convicted of the largest bribery scandal in the history of federal contracting, having collected some $30 million in bribes and kickbacks.

Pipe-dreaming politicians still abound. Some have proposed pumping hundreds of thousands of cubic feet of water from the upper Mississippi River system into the upper Colorado River basin to ensure the water supply for booming Las Vegas, or to recharge the declining Ogallala Aquifer. Another proposal, first suggested in the 1950s, would pipe water all the way from Alaska to the parched Southwest. One group of Alaskan entrepreneurs has even acquired a state permit to export some of Alaska's plentiful and pristine runoff, presumably via converted oil tanker, to industries in Asia.

As Coleridge's shipwrecked Ancient Mariner wisely observed, we actually have "water, water, everywhere, but not a drop to drink." The holy grail to slaking much of our agriculture and urban thirst is the ability to tap the 97 percent of the world's water that now lies salty and unusable just offshore in the oceans' briny deep. No less a dreamer than John F. Kennedy saw the potential a half century ago when he said, "If we could ever competitively at a cheap rate get freshwater from saltwater, that would . . . dwarf any other scientific accomplishment."

The problem for agriculture is one of scale. It takes roughly 2,000 tons of water (1,830 cubic meters) to grow a single ton of wheat. Though the cost of desalination has fallen dramatically in recent years, to about 50 cents per cubic meter, it still makes no sense to buy $900 worth of water to produce about $300 worth of wheat. The cost will certainly fall further as new technologies, such as carbon nanotubes or forward osmosis, come onto the market. But they will likely supply water only to industries or large coastal cities, entities that can afford to pay top dollar. Agriculture, with its high-use and low-value commodities, offers the lowest economic return per gallon.

The key with water, as with many aspects of profligate green-revolution agriculture, is not to waste it. Just as 60 percent of the nitrogen applied to crops either leaches into groundwater or volatilizes into the atmosphere, half of all irrigation water is lost to evaporation and oversaturation of the soil outside of the root zone where plants can actually use it. To put it another way, on average only 50 percent of the water applied to agricultural crops is actually taken up and used by those plants. The other half, along with the tremendous energy required to pump it to the field, is simply wasted, just as half of the water a typical home owner uses to sprinkle the lawn on a sunny day goes right up into the sky.

Peter Gleick, cofounder of the Pacific Institute and author of *The*

World's Water, a biennial report on global freshwater resources, has long argued that we need to take "the soft path" to water sustainability, a term borrowed from energy expert Amory Lovins of the Rocky Mountain Institute, who coined it to argue for increasing energy efficiency instead of energy production. Instead of building more hard structures—dams, canals, pumping stations, and the like—to chase after ever-dwindling supplies, Gleick says, we must increase the productivity and effectiveness of the infrastructure already in place. Conserving water all along the way—lining canals, installing low-flush toilets, utilizing storm water and gray water from sewage treatment plants—makes more and higher-quality water available for home owners, industries, wetlands, and farms, without having to invest billions in costly dams or other new water projects.

There are some places in the world where irrigation can safely expand, and ironically, West Bengal, the site of the 1943 Bengal famine, is one of them. While northern Indian states like Punjab are in the midst of groundwater crisis, the states of southeastern India, such as Bihar, Odisha, and West Bengal, have large underground aquifers that are recharged by monsoonal rains every year. Until recently, poor farmers in West Bengal were prevented from tapping that groundwater by restrictive government policies that were based on the groundwater problems farther north.

Over several years, Aditi Mukherji, a senior researcher at the International Water Management Institute in New Delhi, and her colleagues surveyed 4,000 small farmers in West Bengal, looking at costs and benefits to farmers, as well as various irrigation techniques. Mukherji's research led the government of West Bengal to ease restrictions on tube wells. Unlike in Punjab, where farmers get their electricity for free, in West Bengal water officials put a meter on every pump and charged farmers the same rates for water that urban consumers pay, making it an expensive resource to waste. Right now the government's goal is to connect 100,000 new pump sets to the grid each year. The big beneficiary is the summer, or *boro*, paddy, which under irrigation is already matching or even besting the yields of the typically larger monsoonal rice crop in the fall. Mukherji believes that with widespread access to plentiful groundwater, West Bengal, Assam, and Bihar may become self-sufficient in rice and make a substantial contribution to India's national grain reserve, perhaps taking up some of the slack if water-scarce Punjab's contribution starts to fall.

In the Punjab, farmers are already being forced down the soft path to higher efficiency. The Punjab government is beginning to ration water by reducing the amount of electricity sent to the farmer's pumps each day, and by pushing back planting dates on certain crops. As a result, laser leveling of fields has become almost an obsession in the state, according to Mukherji. By using a laser-guided drag scraper—a common tool for leveling construction sights and roadbeds—farmers can shave their fields to create a 5-millimeter slope from one end of the row to the other. Such leveling costs about $40 a hectare, but it enables the farmer to flood the field 30 percent faster, saving significant energy costs— with 25–30 percent less water in most cases. They even get better nutrient uptake, better weed control, and typically a 10 percent yield increase to boot. The leveling has to be redone every three to five years.

"Farmers everywhere are very innovative species," says Mukherji. "They will find ways of adapting and doing stuff that makes sense. Microirrigation, for example, has been around for 20 years, and a lot of us have been very negative about it, saying that farmers would never adopt it. We were giving 80 percent subsidies, and farmers were not adopting it. Yet just in the last few years we are seeing widespread adoption in places where we had given up. And that's because farmers have just started facing real constraints on water, and that technology started making sense. They will do what they have to do."

Like many other technologies, the cost of microirrigation—a system that uses highly efficient sprinklers or drip systems to apply a precise amount of water to each plant—has plummeted in recent years, thanks to the efforts of people like Peter Frykman. Frykman is a Stanford-trained mechanical engineer who was heading for a lucrative career designing medical devices. While working on his PhD, however, he took a class at Stanford called "Entrepreneurial Design for Extreme Affordability," a lab in which business school students and engineering students worked together to develop products for customers at the bottom of the economic pyramid. As part of the class, he went on a fact-finding mission to Ethiopia.

"I had hardly any concept of what agriculture meant, either on a large scale or a small scale," says Frykman, who grew up in greater Los Angeles. "I had a lot of working ideas, but when we got there it was evident that the primary problem was water. Drip irrigation has been around for 50 years. Why weren't these small-plot farmers using it? We discovered that it was not designed for them. It was too complicated, too expensive. We needed to redesign drip for the other 90 percent of the world's farmers."

Most microirrigation systems are designed for large, well-capitalized farms. They run at high pressures, so that the farthest nozzle from the pump will apply the same amount of water as the one nearest to it. They require a powerful pump and sophisticated sprinklers and valves. But the poor farmers Frykman wanted to help had small fields with rows 100 feet long or less. So he went back to Stanford and designed a low-pressure, non-clogging microirrigation system built from common PVC pipe and polyethylene plastic—the same material used to make cheap plastic bags. He then built a compact laser manufacturing technology that punched precise holes in the polyethylene, which is extruded into rolls of tape about 2 inches wide. He added a plastic rain barrel elevated on concrete blocks to serve as a water tank, a foolproof valve at the end of each row of tape, and voilà!—a cheap, dependable gravity-powered drip irrigation system was born. It can be produced virtually anywhere in the world, can be installed by a local plumber or the farmer himself, and costs $200–$300 per acre: three to five times less than traditional drip irrigation costs, but with the same water savings—typically about half the water used in flood irrigation, with water use efficiency approaching 90 percent. Farmers in China are even using it to apply precise amounts of fertilizer to greenhouse crops—a process known as fertigation—which in some cases can cut fertilizer use in half.

Frykman was so confident in the system's potential that he abandoned his doctorate in 2008 to launch a startup called Driptech. The company has installed the system on 5,000 acres in India, at half an acre to an acre per farm, as well as on several hundred farms in China, and it will soon be manufacturing the system in both countries. Frykman's ultimate goal is to shrink the laser machinery to the size of a standard shipping container, enabling micro-factories to be set up anywhere in the world. If Driptech makes a profit, Frykman believes he can have a larger impact than he could with a nonprofit, and perhaps might also draw the major irrigation manufacturers into the low-end market.

Engineering the system, it turned out, was just half the problem. Selling it to farmers who earn just a few dollars a day was equally challenging. Frykman, who relocated to Pune, India, has been employing a typical Indian marketing campaign. On village market days, his team goes out and advertises through loudspeakers mounted on the back of a pickup. They have a demonstration kit with a 2-liter soda bottle attached to a strip of Driptech tape, and when they turn the bottle over, little fountains of water, all the same height, come shooting out. Even farmers with minimal education get the concept immediately. Creating a product for purchase, instead of as a giveaway from an aid organization, fosters a more respectful relationship, Frykman says. Anyone will take something for free, but if they have to buy it with their limited income, it has to work perfectly, and they will have a vested interest in maintaining it.

"For us, the social impact is inherent in our product," says Frykman. "It's a tool that will make more money for the farmer and his family. People always look at small-plot farmers and say, 'Oh, the problem is lack of education, or lack of health care, or lack of gender equality.' But those are symptoms. The problem is poverty, and if you want to cure poverty you have to help them make more money. And when the poor farmer makes more money, they invest in health and education, and all these other symptoms disappear." Other entrepreneurs agree. Driptech was the first company to receive financing from the Khosla Impact Fund, a venture-capital fund started by billionaire Sun Microsystems cofounder Vinod Khosla to create social impacts around the globe.

Other agronomic water-saving techniques are gaining traction around the world. In one six-year study of irrigation techniques in winter wheat on the North China Plain, Chinese agronomists discovered that if they irrigated their wheat fields only twice a season instead of the typical four times, they saved 25 percent or more water and still got decent yields. Known as "deficit irrigation," the technique requires giving crops less than the full amount of water they need, particularly during growth stages that are less water-dependent, such as the early vegetative stage or late maturing stage. Indian scientists actually increased the production of groundnuts by cutting back on their water, which helped the plants develop deeper root systems to more efficiently use the water they did receive. The technique seems to work particularly well with fruit trees. Australian orchard growers using deficit irrigation found that their water productivity—the yield per amount of water applied—increased 60 percent, with no loss in yield. They even reported higher-quality fruit.

Improving the efficiency of irrigation on 20 percent of the world's farmland will be critical as water demand rises from all corners. But it can't solve the looming problem alone. About 60 percent of our food still comes from low-yielding, nonirrigated farmland that depends on decent rains to make a crop, and the level of production on these farms will be key to solving any future food crisis. Luckily, there are numerous well-known, proven agronomic techniques that help conserve soil moisture and boost yields during droughts. One is no-till farming, in which seeds are sliced or drilled into the previous crop's residue, leaving the soil surface relatively undisturbed. Another is planting cover crops in winter that increase the porosity of soil, helping it retain water like a sponge. Ohio State University's Rattan Lal—one of the preeminent soil scientists in the United States—is a huge fan of plain old mulch, which slows down evaporation, keeps soil temperatures cooler, and reduces water loss. Agroforestry, the technique of growing annual crops under the shade of perennial trees that can provide fuel, produce a food crop, or add nutrients to the soil, is another proven method of getting more crop per drop. Just switching to less thirsty crops, such as sorghum or millet instead of maize, can cut water use significantly.

Farmers in drought-prone developing countries are already boosting their yields through simple supplemental irrigation, catching and storing off-season rainwater in tanks or ponds and then using it to water their crops during periods of high water stress. And despite legitimate concerns about groundwater depletion in many parts of the world, there are also vast untapped groundwater reserves in parts of sub-Saharan Africa and South America that could be carefully and sustainably tapped with the right infrastructure and policies.

Just improving the abysmal performance of the old, often poorly maintained canals and pumps that are already in place would be a good start, says Aditi Mukherji:

"India has made billions of dollars of investment in public irrigation, and farmers basically exited out of those systems because groundwater provided a viable option," she says. "Now that in many parts of the country even groundwater is not viable, I think farmers would have renewed interest in going back to those public irrigation systems. But how does that public irrigation bureaucracy respond to those demands? I think that would be a big challenge. In many cases there is no water in those canals anymore. Those need to be managed and brought back into service. Not necessarily new dams and new construction, but renewing what was already created."

Such work is already under way in the Imperial Valley. In 2003, Lloyd Allen and other southern California water leaders signed the Quantification Settlement Agreement, a landmark water deal that reduced California's overuse of the Colorado River to its allotted share of 4.4 million acre-feet. As part of the agreement, San Diego agreed to pay for lining the All-American Canal, enabling the perennially parched beach town to keep the 22 billion gallons saved from reduced leakage each year. An additional 30 million acre-feet of water will shift from agricultural to urban use through voluntary land fallowing, in which farmers are paid $500 an acre to grow nothing but tumbleweeds.

Moving water from farms to cities is not without impact. Fallowing means fewer farm jobs in the valley, which routinely has the highest unemployment in the state. It also means less drain water flowing into the Salton Sea, the shrinking, hypersaline, often putrid body of water that is California's largest lake and a federally designated sump for agricultural runoff from valley farms. Since 90 percent of California's wetlands have been drained or paved over, the algal soup that is the Salton Sea has become a critical stopover for migratory birds on the Pacific Flyway. More than half of the 800 bird species found in the United States can be seen there, from giant clouds of snow geese, to endangered Yuma clapper rails. Unless the debt-ridden state government spends millions—if not billions—on a restoration plan for the lake, it will eventually become a shallow brine pond surrounded by miles and miles of pesticide-contaminated salt flats.

The death of the Salton Sea doesn't bother Lloyd Allen, who grew up hunting and fishing on its shores. The lake was created by irrigation a century ago when a diversion dam blew out and flooded the dry Salton Sink, which lies more than 200 feet below sea level. It was sustained by irrigation drain water throughout the twentieth century, and it will one day return to the dry Valley of the Dead when farmers are forced to use water more sparingly and the excess is transferred to the state's booming cities. The day we stood by the East Highline Canal, Allen told me we should just be thankful for all the lake has given us and let it go—fish, birds, and all.

"This Salton Sea thing is just a distraction from . . . growing food and fiber for the nation," Allen said. His allegiance was plain. He was a hard-path man. When the Quantification Settlement Agreement was signed with the secretary of the interior in Boulder City next to the Hoover Dam, Allen made headlines by telling reporters he was going to walk out into the middle of the gargantuan structure, get down on his knees, and plant a big fat kiss on the concrete—because it was responsible for making his valley so green and beautiful. "Man, the local papers reamed me good," Allen said. "They thought I was being flip. But I was just being appreciative of what that dam does for us. Without it, there would be no LA, no Las Vegas, and no Imperial Valley."

A few years later I paid my own lip service to California water, and if Lloyd Allen's kiss was an homage to the past, mine was a nod in the opposite direction. While reporting a story on California's devastating drought in 2009, I visited the Orange County Groundwater Replenishment System, a throbbing and humming collection of pumps, pipes, and tanks not far from the original fantasy world of Disneyland. It is one of the largest water reclamation facilities in the world. Each day the giant warehouses full of high-tech membranes and microfilters use reverse osmosis to recycle 70 million gallons of Orange County sewage into water so clean it is almost distilled. There, a polite young engineer named Shivaji Deshmukh handed me a cup of recycled sewage water that likely had been swirling around an Anaheim toilet bowl just 24 hours earlier. I took a gulp. It was cold and crisp and clean with no lingering regrets, guilt, or aftertaste.

It tasted a lot like California's future.

Flood-resistant rice—a product of modern genetic technology—grows in a greenhouse at UC Davis.

CHAPTER 11

Magic Seeds

Feeding Shareholders or the World?

> In my dream I see green, vigorous, high-yielding fields of wheat, rice, maize, sorghums, and millets, which are obtaining, free of expense, 100 kilograms of nitrogen per hectare from nodule-forming, nitrogen-fixing bacteria.—*Norman Borlaug, 1970*

In 2002, several countries across southern Africa found themselves on the verge of famine. Torrential rains the previous year had flooded much of the critical maize crop, and the following season erratic showers dribbled off into the worst drought to bake the region in a decade. Just as during the Bengal famine of 1943, poor government policies exacerbated the natural disaster. Zimbabwe's President Robert Mugabe ramped up his campaign to seize large commercial farms owned by white Zimbabweans, crippling production. At the same time, the government of perpetually malnourished Malawi, in an effort to pay down its national debt, sold all 165,000 tons of its grain reserves, leaving the country with nothing for emergencies. By the spring of 2002, more than 14 million people in Zambia, Zimbabwe, Malawi, Mozambique, Swaziland, and Lesotho were facing starvation. Many resorted to eating leaves and roots, watching helplessly as their children's bellies swelled with hunger.

The World Food Programme of the United Nations leapt into action, requesting a million tons of emergency food aid from donor nations—more than half of which came from the United States. Shiploads of corn, soy, and cornmeal were promptly dispatched to southern Africa. By July the food aid had begun to flow. Everything went smoothly, with distribution centers passing out donated grain to the starving.

Sending food to hungry people sounds simple, but the politics of food aid is anything but. In the midst of the crisis, the leaders of the starving countries discovered that some of the corn donated by the United States came from genetically modified varieties. A year earlier the African Union had approved a model law on biosafety that allowed African nations to severely restrict the flow of GM crops into their countries. To opponents of the controversial technology, the United States appeared to be using the crisis to slip GMOs into Africa. The proverbial "frankenfood" hit the fan.

Swaziland and Lesotho, countries dependent on US trade, took the food aid without much objection. But Malawi, Mozambique, Zambia, and Zimbabwe balked, citing concern for their native biodiversity and their citizens' health, especially HIV sufferers, even though thousands had eaten the grain without ill effects. The African leaders were also afraid that farmers would save the free seed and plant it, jeopardizing critical agricultural trade with Europe, where GMOs must be segregated and labeled—a prohibitively expensive task for poor nations with millions of small farmers. As a result, these countries halted distribution of free food in the midst of the famine, and the red-white-and-blue bags of US grain piled up in guarded warehouses.

Malawi, Mozambique, and Zimbabwe eventually accepted the proffered corn on the condition that it be milled first to prevent it from being planted. But Zambia held up the US grain until a panel of the country's top scientists deemed it safe. The fact-finding mission visited the United States and South Africa, where GM crops are widely grown, and several countries in the European Union, where opposition to them is intense. Though 3 million Zambians were at risk of starvation, the panel sided with Europe, concluding that the donated maize threatened local varieties and future exports, and that the health issues remained unresolved. One British activist even told Zambia's *Daily Mail* newspaper that the virus used to create GMOs "could form a retrovirus that could produce symptoms similar to HIV"—the dietary equivalent of shouting "Fire!" in a crowded theater.

The rhetoric from US officials was just as vitriolic, with one quoted as saying "beggars can't be choosers." Even Andrew Natsios, then head of USAID, claimed that anti-GMO groups were "putting millions of lives at risk in a despicable way."

The Zambians found enough non-GM corn from other nations to tide them over until their next harvest, and few famine-related deaths occurred. To this day, an effective ban on GMO crops stands in many of Africa's hungriest countries, as well as most of the European Union.

FEW ISSUES HAVE created a bigger global food fight than the prospect of eating plants or animals whose DNA has been altered in a petri dish. Yet in some respects, the GMO debate is just the latest round in the long-simmering controversy over the green revolution. GMOs raise many of the same concerns as persistent pesticides like DDT, including the potential for widespread effects on wildlife and human health, as well as fears of growing corporate control of agriculture.

Hyperbolic rhetoric predominates. Pro-GM lobbyists and advocacy groups, heavily funded by the multi-billion-dollar biotech industry, characterize their opponents as well-fed, technophobic food snobs who want to impose pest-ridden agriculture on a starving world. The anti-GMO argument, put forward by some well-funded environmental groups, paints an equally ludicrous picture of unnatural frankenseeds that put human health and biodiversity at risk so that corporate agricultural giants can control the world's seeds and enslave poor farmers in the name of capitalist greed and profit.

The truth, as always, is somewhere in the middle.

First, a bit of objectivity. Unless you are a feral wilderness dweller living on game, fish, and berries, virtually everything you eat has been genetically modified. Most of the genetic engineering was accomplished by our ancient ancestors who first domesticated crops, and the early farmers that followed. By saving seeds from the tastiest, highest-yielding, and best-looking plants over the millennia, they winnowed out unwanted genes. The same holds true for livestock and even pets. The reason I have a sweet yellow Labrador snoring at my feet instead of a hungry she-wolf is that "Pearl" is a highly specialized GMO, carefully selected, crossbred, and then heavily inbred by early humans and later northern hunters to lick my hand, guard my family, and instinctively retrieve any duck I might knock from the sky.

During a recent study of the genome of modern corn—the most highly bred staple crop on the planet—researchers at UC Davis found that early Meso-American farmers changed the plant's genetic makeup far more drastically than modern plant breeders have. Even Norman Borlaug's historic transformation of wheat from tall, wispy, reedlike plants to short, stocky ones involved the transfer of a just a few genes. But it took him two decades and thousands of crosses to find them.

IT WASN'T UNTIL the 1970s, when molecular biologists developed recombinant DNA technology, that humanity seemed to cross a scientific and moral Rubicon. Instead of simple crossbreeding or exposing seeds to mutagenic chemicals or gamma rays in hopes of generating beneficial mutations (Borlaug's only hope for developing nitrogen-fixing grain in 1970), we could now actually insert slivers of genetic material from one species into another—creating hybrids that would never occur in the wild. Bacteria could now, essentially, mate with plants. Fish could pass genes to tomatoes. An entire sheep could be cloned from the mammary tissue of another. With the development of genetic markers, molecular cloning, and other tools, scientists began unraveling the black box of the double helix. The "gene revolution" was on.

Plant breeders were beside themselves with this transformational new technology. I remember listening to my plant-breeding professor at North Carolina State wax on about transgenic crops. Before long, he predicted, we would be able to eliminate nutrient deficiencies around the globe with fortified grains, saving millions from preventable disease. We would be able to vaccinate poor children with fruits and vegetables that not only contained the vaccines, but produced them as well. We would finally be able to fulfill Borlaug's dream and genetically shift nitrogen-fixing rhizomes from legumes onto corn plants and never have to apply nitrogen fertilizer again. It would be good for the planet and good for poor farmers who couldn't afford fertilizer. It sounded good to me.

Three decades later none of those goals have been met. What

we've received instead are genetically engineered crops that have
been far more beneficial to biotech companies and large commercial farmers than to malnourished children or subsistence farmers. Unlike the green revolution, which emerged from foundations
and land-grant universities, the gene revolution was born largely
in the labs of chemical and pharmaceutical industry giants like
Monsanto, DuPont, Syngenta, and Bayer. These companies rapidly bought small, family-owned seed companies during the 1980s
and 1990s to gain access to a wide range of genetic material. By
2008 the four largest seed companies controlled 56 percent of all
commercially branded seeds. Not surprisingly, the first products
they made were, like their other agrochemical products, marketed
to large-scale, relatively wealthy farmers who could afford the
hefty prices the companies had to charge to recoup their R&D
investment.

The first significant agricultural product to come from the gene
revolution was recombinant bovine somatotropin (rBST), also
known as recombinant bovine growth hormone (rBGH), marketed as Prosilac. Monsanto scientists found that if they inserted
a strand of cow DNA into a certain bacterium, it would pump out
high levels of a naturally occurring growth hormone, somatotropin, in cows. When injected twice a week into dairy herds, Prosilac raised milk production by about a gallon per day per cow, or
roughly 15 percent. That put more milk money in farmer's pockets
from the same herd, but it led to more udder infections, decreased
fertility, increased lameness, and more antibiotics for the cows. At
the time, the nation did not need more milk. The USDA had just
spent nearly $2 billion in the late 1980s buying out 14,000 dairy
farmers (who slaughtered or exported 1.6 million dairy cows) to
reduce milk surpluses. Nonetheless, the FDA approved Prosilac's
use in dairy cows in 1993 after studies found milk from treated
cows identical to that of nontreated cows. Canada, the European

Union, New Zealand, and Australia, however, immediately banned the hormone on animal welfare grounds.

In the United States, doping beef cattle with steroids and other growth hormones has been a common practice since the mid-1950s. By 2002, about 22 percent of the US dairy herd of 9 million cows was on Prosilac. But the consumer backlash was intense. Companies like Ben & Jerry's ice cream and Tillamook cheese quickly eschewed dairy products from such herds, and Starbucks, Wal-Mart, and Costco eventually followed suit. Opposition to the drug actually kick-started the organic dairy industry, which grew from 2,000 cows in 1992 to more than a quarter million by 2011. By 2010, less than 10 percent of US dairy cows were still on the genetically altered drug.

Despite the controversy, Prosilac was still a niche product. But shortly after it won FDA approval, Monsanto developed the first blockbuster GMO. Company scientists inserted a gene from a common soil bacterium into a soybean variety that made the resulting plant impervious to Roundup, one of the company's best-known herbicides. Glyphosate, Roundup's active ingredient, had been around since 1974, but it had limited agricultural use because it killed every green thing it touched, including crops.

With Roundup Ready soybeans, however, farmers could now spray their entire fields with one chemical, kill the weeds, and not harm a single soybean plant. The seeds cost about 40 percent more than conventional varieties, and farmers had to sign legally binding agreements that forbade saving any of the crop for replanting the next year—a tradition as old as agriculture itself. The Roundup Ready transgene was deemed by law the patented intellectual property of Monsanto. Even so, commercial soybean farmers readily adopted Roundup Ready seeds, because they saved fuel, time, and money, largely by making it easier to plant "no-till."

Instead of the traditional method of plowing or disking their fields in the spring and spraying a cocktail of chemicals to kill weeds, farmers could now simply spray the fields with Roundup. They could then use a specially designed planter that slices the seeds directly into the stubble, keeping the soil surface largely intact. The practice has been promoted for decades for erosion control, but before the Roundup-resistant trait it was never easy to do, because the weed pressure on most fields is intense. Now, with the new soybeans enabling weed control with just a few applications of glyphosate, and no-till planters cutting out tillage, farmers reduced the number of trips across the field from seven or more each season to four or fewer. Making fewer trips cut their fuel use by 50–80 percent and their labor by 30–50 percent over conventional soybeans, lowering their costs and leaving more money in their pockets. Monsanto soon added the Roundup resistance trait to corn and cotton seeds. Sales went through the roof, as did the use of glyphosate, which soared from less than 5,000 metric tons in 1987 to more than 80,000 metric tons in 2007 in the United States alone. It soon became, and remains, the best-selling pesticide in the world.

THE COMPANY FOLLOWED with a second genetically modified trait that has become nearly as widespread—a transgene that causes corn, cotton, and potato plants to produce toxic proteins from the common soil bacterium *Bacillus thuringiensis*, better known by its initials, Bt. This fascinating bacterium produces dozens of proteins that have no mammalian toxicity, and in most cases won't even harm beneficial species like earthworms, honeybees, or ladybugs. But when eaten by certain lepidopteran insects—such as corn rootworms or European corn borers—Bt dissolves their digestive tracts, killing the bugs in a few days. Organic farmers have used Bt sprays on their crops for decades; Rachel Carson even proposed it as an alternative to toxic organophosphate insecticides like DDT.

It's now been two decades since these two traits were approved, and no other blockbuster GMO has followed. The vast majority of all GMOs grown in the world today are corn, soy, or cotton plants that carry one or both of these traits. Despite the extreme controversy they've caused, they have provided legitimate environmental benefits. By 2012 the amount of US farmland in no-till had surged to 92 million acres—roughly a third of US fields—significantly cutting soil erosion, fuel consumption, and greenhouse gas emissions.

Glyphosate, the active ingredient in Roundup, has a very low mammalian toxicity and was long thought to break down quickly into molecules that bind tightly to soil particles, posing little risk of runoff or groundwater contamination. This was a vast improvement over the previous herbicides used in no-till, such as the acutely toxic paraquat and the highly mobile atrazine. The latter, still widely used in cornfields, is a known endocrine disrupter linked to heart disease in humans and sexual abnormalities in amphibians, fish, reptiles, and even mammals. It is a ubiquitous contaminant in lakes, streams, and aquifers in rural America, and it is one of several agrochemicals that pediatric researchers suspect of causing the annual bump in birth defects in US babies conceived between April and July, when most pesticides are sprayed. Similarly, cotton is among the most heavily sprayed crops on the planet, and the insecticides used—as the farmers from the Punjab have sadly learned—are among the most acutely toxic agricultural chemicals on the market. Bt crops in 2011 alone cut the quantity of insecticide sprayed on the world's cotton fields by an estimated 37 percent (amounting to a reduction of 17 million kilograms of active ingredient). Much of that reduction was among small farmers in India and China. The emissions savings from both GM traits amount to removing 840,000 cars from the road each year.

So far, the environmental and health nightmares that many anti-GMO groups feared have yet to materialize. After reviewing hundreds of studies, the National Academy of Sciences in the

United States and the European Union's Directorate-General for Research and Innovation determined that current GMO crops pose no new risks to humans or the environment. That view has been endorsed by the World Health Organization, as well as by the UN's Food and Agriculture Organization, the latter claiming with uncharacteristic definitiveness that "no verifiable untoward toxic or nutritionally deleterious effects resulting from the consumption of foods derived from genetically modified foods have been discovered anywhere in the world."

Many nations, consumers, and environmental organizations remain unconvinced. Most of those safety claims are based on industry studies conducted to win regulatory approval, which are not available for peer review by independent scientists because the data are deemed proprietary. When independent researchers have reported harmful effects from GMO crops, they have been met with a hailstorm of criticism and counterstudies from industry and pro-GMO academics—a response that smacks more of PR damage control than serious scientific debate. The dissenting researcher is typically vilified in the process, doing little to assuage public concerns.

Early studies did ring alarm bells. Dr. Arpad Pusztai, a renowned researcher at the Rowett Institute of Nutrition and Health in Scotland, found harmful effects on rats fed transgenic potatoes that had been engineered to produce lectins, plant proteins that can be potent insecticides. After sharing his concerns in a BBC interview in 1998, Pusztai was promptly fired from the institute, where he'd worked for 40 years. The potatoes were never produced. The first version of Bt corn produced pollen that actually did harm monarch butterflies, though the offending protein was quickly replaced. And in 2000, a transgenic corn variety named StarLink found its way into the taco shells of Taco Bell, even though it had not been approved for human consumption. StarLink caused no illness and was eventually given the green light for humans to eat, but the mix-up showed just how hard it is to control GMOs in farmers' fields and in the giant maw of the US industrial food system.

Studies also continue to surface suggesting that glyphosate may not be as benign as was once thought, which is particularly troubling, given that the EPA has allowed significant increases in glyphosate residues in food since Roundup Ready varieties were introduced. Contrary to early beliefs, government researchers in 2011 surveyed 38 states and found low levels of glyphosate and its breakdown products to be widespread in US groundwater, surface water, soils, and even rain. Research from South America, published in the peer-reviewed journal *Chemical Research in Toxicology* in 2010, suggested that the chemical actually did cause birth defects in frogs, rats, and chickens. One Canadian study, published in another prestigious journal, found that older pregnant mothers exposed to glyphosate had three times the risk of spontaneous abortion compared to unexposed mothers of similar age. Other researchers have reported a link between maternal exposure to glyphosate and neural-tube defects.

Bt crops recently raised concerns as well. In two large, replicated feeding trials, researchers at the Swiss Federal Institute of Technology in Zurich found that pollen from Bt maize plants increased the mortality of ladybird beetles, an important beneficial insect predator that feeds on crop pests. As usual, counter-studies and criticisms ensued, but the Swiss study was deemed credible enough by German food authorities, who cited it in their subsequent ban of Bt maize in Germany.

Although such worrisome studies are still outliers of the scientific consensus, GMO critics are dead right about one thing: the widespread use of Bt crops and glyphosate has accelerated the rise of resistant insects and weeds. Neither Monsanto nor the USDA required farmers to rotate Roundup Ready varieties with non-

GMO crops or to plant buffer zones to thwart this workaround of nature. They did require farmers planting Bt crops to provide 20 percent of their acreage as non-Bt refuges, and up to 50 percent for some varieties of Bt cotton. But a 2009 study found that nearly one out of every four corn farmers was ignoring the law. As corn prices rose to record levels through the first decade of the twenty-first century, many farmers abandoned crop rotation—perhaps the single most important agronomic practice—and began planting GM corn season after season.

They are now reaping what they sowed. By 2012, more than 61 million acres of US cropland was infested with glyphosate-resistant weeds, almost double the area just two years earlier. Nearly half of US farmers surveyed in 31 states had such superweeds in their fields. The problem was so bad that Monsanto began offering farmers rebates of up to $12 per acre to buy older, more toxic pesticides, including competitor's products, to kill the plants. Bt-resistant bollworms are now happily munching on cotton fields in India, China, and the United States, while Bt-resistant corn rootworms are popping up in the US Midwest, requiring biotech companies to stack new Bt proteins in their seeds. Dr. Michael Owen, an agronomy professor at Iowa State University and a member of the National Research Council, warned a congressional committee in 2010 that, if not managed properly, the rapid rise of glyphosate-resistant weeds could cancel all the environmental benefits of Roundup Ready crops, particularly if farmers are forced to give up no-till farming and return to disking their fields.

In a disturbing turn of events, the biotech industry is now rushing to develop GM crops that also resist older and far more toxic broadleaf herbicides. These include dicamba (invented in 1961) and the granddaddy of them all, 2,4-D, which entered the US market in 1945 and was a major ingredient in Agent Orange, the defoliant used during the Vietnam War that is now blamed for sickening US veterans and causing tens of thousands of birth defects in Vietnamese children.

Indeed, chemical companies have invested so much of their R&D dollars in genetically engineered seeds that the development of newer, safer pesticides has all but dropped off their radar. No major new herbicidal mode of action—the way a chemical affects plant tissues or cells that typically defines an entire class of herbicides—has been registered in the last two decades. Nor has the biotech industry developed any more blockbuster GMOs with widespread benefits for the world's farmers or the environment. During a 2009 interview with Robert Fraley, the chief technology officer for Monsanto, widely regarded as the "father of biotechnology," I asked when I was going to see Borlaug's nitrogen-fixing corn on the market. "I don't know about nitrogen-fixing corn for a while," Fraley said.

"IF YOU'RE WAITING on private companies to do research in the public interest, it's never going to happen," says noted agricultural economist Bruce Babcock at Iowa State. "There is tremendous investment in technology for crops in which they can capture the profit. It's what we call capitalism. Who can pay for a $200 bag of seed corn—a poor African farmer or an Iowa corn farmer? Now if Monsanto can sell the African a $50 bag of corn and make a profit, they'll do it."

Biotech companies are developing one new trait aimed at helping farmers adapt to climate change: drought-tolerant maize. Monsanto has been promising to deliver this highly hyped GMO to farmers since at least 2009, and to share the genetic material—royalty-free—with researchers in Africa. DuPont and Syngenta developed non-GMO drought-tolerant seeds that are already on the market. Several Midwest corn farmers tested the seeds in 2012, during the worst drought to hit the Corn Belt in decades, and were generally impressed. Many were surprised that they harvested any corn at all.

Dr. Qingwu Xue, a plant physiologist at Texas A&M University, has conducted the most authoritative comparison of the new drought-tolerant seeds in the scorching Texas panhandle outside Amarillo. The region has some of the highest corn yields in the nation, with farmers averaging 200 bushels per acre in most years. All of those fields are irrigated from the dwindling Ogallala/High Plains Aquifer, which is falling faster in Texas than in any other state. Xue watered each variety at four different rates from 100 percent of the corn's water demand to 40 percent, mimicking drought conditions. The drought-tolerant seeds were compared to the top hybrid varieties currently grown in the region.

"At full irrigation potential, I don't see much penalty with conventional varieties," Xue says. "But less than full irrigation, we do see a yield difference of up to 20–30 bushels per acre." Most of that difference came at the lowest water levels. The upshot for Xue was water conservation. Farmers using drought-tolerant varieties could apply only 75 percent of the corn's water needs and still get close to 200-bushel-per-acre yields. That amounts to 5 inches less water per acre, and each inch of water not applied saves 1.3 billion gallons of precious fossil water per acre each year, extending the life of the Ogallala.

DuPont's and Syngenta's non-GMO drought-tolerant varieties performed about the same as Monsanto's transgenic corn— leading critics to argue that GMOs are unnecessary. Xue, like most agronomists, is pragmatic. "I think everything is helpful," he says. "The real dilemma for us is that you want to keep yield potential increasing, but the stress levels are rising because of climate change. We see more natural disasters and their scale is ever bigger. GMOs are just a new tool. If you relied on natural selection, I don't know how you could feed 9 billion people."

BUT CAN GMOS actually deliver on that promise to create wide-spread environmental, nutritional, and yield benefits for the world's poorest in the face of climate change? The answer is a resounding . . . maybe—particularly in the crop that feeds half the world: rice.

In July of 2000, Ingo Potrykus, a visionary plant geneticist, was pictured on the cover of *Time* magazine staring out past saffron-specked tendrils of rice heads. Over the previous decade, Potrykus, a professor at the Swiss Federal Institute of Technology in Zurich, and his colleague Peter Beyer at the University of Freiburg, Germany, discovered that if they inserted two genes from the daffodil flower into rice, the rice would produce beta-carotene, the substance that our bodies convert to the critical micronutrient vitamin A and that is found in orange sweet potatoes, carrots, spinach, and meat.

The process was harder than it sounds. Though rice plants produce beta-carotene in their leaves and husks, the compound doesn't make it into the endosperm of the rice that consumers eat. Beyer and Potrykus had to genetically reengineer a new biochemical pathway in their rice. Early varieties carried little beta-carotene, but with help from scientists at Syngenta (who swapped the daffodil gene with a more effective one from maize), by 2005 there was enough in a single 100-gram serving of golden rice to provide the bulk of the daily vitamin A recommended for children and pregnant mothers by the World Health Organization. It also came with a nice golden hue.

Vitamin A deficiency has long been one of the greatest scourges of the developing world, crippling the immune systems of pregnant mothers and an estimated 40 percent of the children living in those regions. It is a major contributor to blindness in some 500,000 children every year. The World Health Organization estimates that 250 million preschool-aged children suffer from vitamin A deficiency. A reliable, affordable beta-carotene source could prevent a third of the deaths of children under five every year— roughly 2.7 million kids.

Nations and nonprofit groups have known about this problem for decades. Programs to fortify processed foods, distribute vitamin pills to the poor, and promote more balanced diets work well in cities but are less effective in the rural hinterlands, where 45 percent of the vitamin A–deficient population lives. There, balanced diets are expensive and hard to come by, and the bulk of people's calories come from cheap white rice. Current vitamin A pill programs cost an estimated $500 million each year. The distribution of Golden Rice seeds, on the other hand, could cost as little as $30 million and generate an estimated $12 billion a year in economic benefits from less disease and better health. Syngenta, Novartis, Bayer, Monsanto, and other biotech firms agreed to donate some 70 patents involved in the development of the rice to a nonprofit foundation that would manage the variety. Farmers will get the seeds for free and can save their seeds to replant year after year.

GMO critics, however, see the fortified rice as the ultimate Trojan horse for the biotech industry. Greenpeace, with its staunch, no-exceptions advocacy against genetically modified foods has campaigned tirelessly to keep Golden Rice out of the rice-dependent countries of the world, where vitamin A deficiency is most acute, and so far the organization has been successful. Greenpeace has been aided in its cause by a scandal surrounding a Tufts University feeding trial that tested the bioavailability of beta-carotene in US adults and 25 Chinese children who ate Golden Rice. Though no children showed adverse reactions and the rice provided 60 percent of their recommended intake of vitamin A, the researchers failed to inform the participants or their parents that the rice was genetically modified—a clear breach of typical FDA protocols regarding informed consent. More than 30 researchers from around the globe sent a letter of protest to Tufts claiming that the trial violated the Nuremberg Code—the guiding research principles established after World War II to prevent the kind of horrid human experiments conducted in Nazi concentration camps.

Golden Rice may be losing its stigma, however. The Philippines, home to the International Rice Research Institute, where much of the Golden Rice development took place, will likely be the first nation to approve the new GMO for planting in farmers' fields. If it yields well and tastes good—perhaps the most important test of any new variety—it could launch a Golden Rice revolution in Asia. Already researchers in Australia are following suit, hoping to produce a beta-carotene-fortified "Golden Banana" for Africa by the end of the decade.

It won't be soon enough for Ingo Potrykus, who has written that costly and restrictive regulation of GMOs based on ideology, instead of actual risks and benefits, hinders public research on crops geared to the greater good, leaving the technology in the hands of a few wealthy companies that see no profit in developing crops for the poor. Such rules, he argues, constitute nothing short of a "crime against humanity."

Luckily, the cost of genetic research has fallen dramatically since the first plant genome was sequenced in 2000, enabling more public researchers to jump into the GMO game. "To sequence the *Arabidopsis* genome—the fruit fly of plant breeders—took $70 million and seven years," says Pam Ronald, a rice geneticist at UC Davis. "The same process is now done in a week for $99. You can now do DNA sequencing on computers not much more powerful than your cell phone."

The technology was critical in enabling Ronald, along with her colleague David Mackill at IRRI, to develop a new rice variety that could hold its breath underwater. Though rice loves to keep its feet wet in shallow, flooded paddies, if the entire plant is submerged it will die in just a few days. Some 50 million acres of rice is planted in flood-prone areas, and sea level rise is expected to make floods more frequent. Rice researchers in the 1950s discovered a wild rice variety that seemed to withstand flooding. Its yield was pathetic, but something in it kept it alive underwater. Traditional plant breeders at IRRI spent the next few decades crossing the wild rice with conventional varieties in an attempt to breed the trait into something farmers could grow. But each time the trait was crossed, it brought its low-yield baggage with it, bearing hardly enough seeds to produce the next generation. Researchers finally put it on a shelf and forgot about it.

By the mid-1990s, with the gene revolution in full swing, Ronald and Mackill turned back to the submarine rice. Using genetic markers and DNA sequencing, they found the trait they were looking for on chromosome 9. But chromosome 9 was long and did a lot of things. They winnowed the trait's location down to 20 genes, then 10, and eventually 1, that they and colleagues at IRRI were able to transfer successfully into a popular cultivar called Swarna. Indian farmers planted the first "scuba" rice in 2008. It survived floods so well that farmers eagerly shared the seeds. By 2013, more than 5 million people throughout Asia were growing the flood-resistant rice, and its acreage is expanding every year.

The project was groundbreaking for the future of plant genetics. Both critics of GMOs and some plant breeders had long argued that no single gene could control such a complex agronomic trait. More likely, a suite of genes was responsible, and they were probably scattered all over the plant's chromosome. Tiny needles in a microscopic haystack. Yet Ronald and Mackill found one gene on one chromosome that did the trick. Though the new rice was not technically a GMO, since the gene came from another rice plant, the researchers used all the GMO technology at their disposal to find the gene and transfer it into a successful cultivar without any baggage.

Ronald and Mackill have won several prizes for their work. Unlike the development of glyphosate-resistant corn, or Bt cotton, their research is publicly funded, and free and available to all. Such work would never have been done by a private company, says Ronald, which is why public research remains so critical. "If you want to improve a poor farmer's food security," she says, echoing Norman Borlaug, "the best thing I can do as a scientist is supply them with improved seeds."

Flood tolerance, like drought tolerance, is a good thing. But it still affects only the rice growers who live in flood-prone areas. Researchers in Australia are also working on salt-tolerant rice, for planting in areas where overirrigation has left soils too salty to plant in, as well as on crops that utilize nitrogen and phosphorus more efficiently. With the new genetic tools, plant breeders may finally be able to make plants resistant to weeds themselves, or create crops that repel insect pests instead of killing them.

All these things are helpful around the margins. But they are still largely mitigation strategies to help farmers maintain the world's grain production in the face of increasing stress from climate change. None of them promises to double yields as the green revolution did, save one. Known as C_4 rice, it is the rice breeder's holy grail.

All plants on the planet convert sunlight and carbon dioxide into the sugars they need to grow in one of three ways. The first path, crassulacean acid metabolism, or CAM, is common among desert plants and not much use to farmers or crops. A staggering 98 percent of all plants on Earth get their energy via a second ancestral pathway in which chloroplasts produce a three-carbon molecule that is broken down into plant food. Biologists call the process the "C_3 pathway." The advantage of C_3 plants is that they typically become more efficient as atmospheric CO_2 rises. That's why many scientists initially thought crop yields would rise from all the extra carbon in the atmosphere. The downside of C_3 plants, however, is that, as temperatures rise, they lose more and more carbon energy through photorespiration at night—which is largely why researchers have found no actual yield boost.

But some 50 plants, mostly grasses and sedges, have evolved a third, newer, highly efficient pathway that produces a four-carbon molecule. These "C_4 plants," including corn, sorghum, and sugarcane, are among the most efficient and highly productive plants on the planet. The C_4 pathway is one reason that US corn farmers average nearly 10 tons of grain per hectare, while rice farmers average little more than 4.

Modern rice plants, however, already have many of the structures needed to form the C_4 pathway, and plant breeders have long believed that if they could shift rice over to the C_4 route, they would be able to achieve 30–50 percent higher yields. Even with all the new genetic tools at their disposal, it's a daunting challenge, requiring a significant redesign of the plant's cellular architecture and biochemistry. For years, Jane Langdale, a plant geneticist who heads Oxford University's Department of Plant Sciences and who has studied the genes that control carbon pathways in plants, thought it was a pipe dream. But not anymore. In 2012 she joined a global consortium of plant scientists attempting to do just that. I asked her what had changed her mind.

"The bottom line is that technological advances—particularly in terms of high-throughput [gene] sequencing—gave us a quantum leap in terms of what could be considered feasible," Langdale replied. "I am definitely more optimistic than I was. Progress has been good so far." Best of all, Langdale said, C_4 rice wouldn't need any more fertilizer or water to achieve the higher yields.

The possibilities are mind-boggling. Experts estimate that the world will need another 100 million tons of rice for every additional billion people on the planet. Global rice production in 2012 was about 730 million tons. By that metric, if researchers could increase rice yields by 30 percent, the bulk of the world's food problem might be solved, or at least the status quo maintained, in the rice-eating areas of the world. Still, Langdale estimates it will take 15–20 years of further research to get C_4 rice into farmers' hands.

And that's just the beginning. Another international panel of university scientists recently received a lump of funding to begin research in earnest on developing nitrogen-fixing cereals such as wheat and maize. Four decades after Norman Borlaug's Nobel lecture, scientists are finally getting to work putting rhizomes on corn plants.

We are tinkering with nature, as we have since farming began, and the potential for opening a Pandora's box that we cannot control is ever present as we develop new species that can spread their pollen in the wind. But the Pandora's box of climate change is already open and upon us. It would be the height of foolishness to toss away any tool that has the potential—even if remote—of solving some of the problems that global agriculture has helped create.

MARK LYNAS, A British journalist, author, and outspoken founder of the anti-GMO movement in Europe, has even done an about-face, apologizing for his previous opposition to GMOs and telling the *Guardian* newspaper, "The first generation of GM crops were suspect, I believed then, but the case for continued opposition to new generations—which provide life-saving vitamins for starving people—is no longer justifiable. You cannot call yourself a humanitarian and be opposed to GM crops today."

But there is an important caveat: Not all GMOs are created equal, and each must be judged on its own merits. Corporate biotech companies, whose annual agricultural research budgets have swelled 20-fold in the last three decades, to more than $2 billion a year in the United States alone, remain focused on their bottom line: developing crops that sell more seeds, more pesticides, and create more profit, while aggressively guarding their patents and restricting the flow of their genetic material to public researchers. In 2006 the USDA reported that private companies spent $5 billion on research into food processing and crop development and effectively zero on research relating to the environment and natural resources or human nutrition and food safety. These critical issues, along with basic agricultural research, long-term plant breeding, and germplasm development, remain the domain of government-funded research and land-grant universities that were founded to create a greater public good.

The classic example is Hawaii's disease-resistant Rainbow Papaya, a GMO developed by USDA and University of Hawaii researchers that saved the state's papaya industry from being wiped out by the papaya ringspot virus in the late 1990s. Biotech corporations wouldn't touch it, because the papaya market was too small. Yet federal funding is dwindling, forcing many university researchers to take handouts from big biotech companies that often come with big strings attached. Just as with pharmaceutical research, university investigators who work on the private dime often can't publish their findings if the results don't go the way the company likes. The simplest thing Big Biotech could do to ease public fears of its products would be to publish the health, environmental, and toxicity studies conducted for FDA and EPA approval in peer-reviewed journals that could be replicated and verified by independent researchers. But since that route would require sharing newly developed seeds bearing traits potentially worth billions, it is unlikely.

The public-private partnership that developed Golden Rice remains the exception to the rule, though it could be a model for future collaborations. Much of the current research to feed the world is being funded, as the green revolution was, by land-grant universities and philanthropists, such as the Ford, Rockefeller, and Bill & Melinda Gates Foundations, which helped launch both Golden Rice and submarine rice. At the moment, corporate GMOs are still largely feeding livestock and company shareholders. Researchers funded by taxpayers and nonprofits remain, as they've been since Elvin Stakeman's day, the ones trying to feed the world.

Rodale Institute farm manager Jeff Moyer shows off a cover crop of hairy vetch on the longest-running field trial between organic and conventional farming systems in the United States.

CHAPTER 12

Organic Agriculture

Feeding the Rich or Enriching the Poor?

> The slow poisoning of the life of the soil by artificial manures is one of the greatest calamities which has befallen agriculture and mankind.—*Sir Albert Howard, 1940*

The Grammy Awards have always been a window into American pop culture. The fifty-fourth Grammies, held in February 2012, were no exception. Young British pop diva Adele Adkins won six golden gramophones, while nearly 40 million viewers tuned in to hear Jennifer Hudson pay homage to the late Whitney Houston, who had died of a drug overdose the day before. It was the largest number of viewers since 1984, when the king of pop himself, Michael Jackson, performed his iconic hit "Thriller" to a packed Los Angeles audience.

That night in 2012, a surprise winner emerged, though he didn't utter a sound: a little Claymation pig farmer who appeared in a two-minute ad for the burrito chain Chipotle. As Willie Nelson, a longtime advocate of family farmers, croons a haunting acoustic version of "The Scientist," by Coldplay, the little farmer begins building barns around his pigs, popping them full of pills and pumping out green goo into his ponds as roads and trucks and factories grow around him. His face clouds as he contemplates what he's done in the midst of a bleak winter before he decides to go "back to the start," tearing down his barns and setting his pigs free as the spring sunshine glows, inspiring his neighbors to do the same. The video concludes with the farmer putting a box of presumably happy, healthy pig meat into a Chipotle truck, ending on a farm sign that admonishes us to "Cultivate a better world."

It was enough to make a redneck like me, who can remember how hogs were once raised in the South, to wince a bit at what we've become, with our sterile farrowing houses, caged sows that can barely move, and the 29 million pounds of antibiotics we feed our confined livestock each year just to keep them healthy.

Willie and the cartoon pig farmer struck a chord that resonates deeply (the video has been downloaded 8.4 million times): a growing desire for food grown without harming the planet, animals, or ourselves. This desire manifests itself in the phenomenal growth in sales of foods grown without chemical fertilizers, pesticides, or GMOs, and livestock raised without hormones, antibiotics, or CAFOs. Chipotle, which claims to use "food with integrity," has capitalized on that sentiment. The company uses meats from animals raised without antibiotics or hormones "whenever possible" and uses organic produce "when practical," according to its advertising. The phrasing is a subtle acknowledgment that there's not enough "food with integrity" to supply all Chipotle restaurants around the nation. But even such a nonbinding nod toward a kinder, gentler agriculture—along with pretty tasty burritos—helped the company's sales grow more than 23 percent in 2011 alone, besting nearly every other chain in the country that year.

Organic foods were once the near-exclusive fare of ponytailed, tie-dyed back-to-the-landers and liberal foodies who endeavored to grow their own food or prowled the aisles of co-ops in places like Portland, the Berkshires, or Boulder. But in just two short decades, the organic movement has matured into the fastest-growing segment of the world's food market. Since 1990, organic food sales in the United States have grown by double digits nearly every year. By 2012 the US organic market was worth $28 billion—more than double Monsanto's sales that year. That's still only 4 percent of total food sales in the country, but it continues to grow at a steady clip, while nonorganic food sales—largely tied to population growth—are far flatter, increasing only about 1 percent a year.

Today, nearly 12 percent of fruits and vegetables sold in the United States are organic, along with nearly 6 percent of milk. Though nutritionists and food scientists have found little differences in the nutrition provided by organic foods (they have yet to be found statistically or scientifically "better for you"), they do contain far fewer residues of pesticides, hormones, steroids, and antibiotics. Now co-ops compete with hip, upscale chains like Whole Foods and Trader Joe's that cater to both the college crowd and affluent suburban moms willing to pay a significant premium for "food with integrity."

Such prices are a boon to organic farmers and don't dissuade consumers who have the extra income to spend on organic food. But the growth in organic has made little difference to those farther down the economic food chain. One of the biggest criticisms of organic food is that it is essentially feeding the well-educated and well-heeled, while leaving the poor to get their cheap calories from fast-food chains and convenience stores.

The problem of access to organic foods is slowly fading, however, as big-box discount stores like Wal-Mart and Costco jump aboard the organic bandwagon and the price premiums between organic and nonorganic foods start to shrink. Most organic products are now bought in conventional grocery stores instead of specialty stores, while organic-leaning grocers like Whole Foods now take food stamps. Urban gardens, farmer's markets, and community supported agriculture (CSA) operations are among the fastest-growing food trends in inner cities, from Detroit to Washington, DC, to Oakland. Though studies vary widely, some show that the majority of organic consumers earn less than $50,000 a year.

Yet for many die-hard green revolutionaries and agricultural researchers, the perennial warning about organic agriculture holds sway: great for your garden, but it'll never feed the world. In 2002, Norman Borlaug articulated a common sentiment among agronomists: "We aren't going to feed six billion people with organic fertilizer. If we tried to do it, we would level most of our forest and many of those lands would be productive only for a short period of time." For Borlaug, as well as my professors at North Carolina State, going organic was tantamount to farming the way our great-grandfathers farmed—when yields were a fraction of what they are today and half the world was hungry, even with a third fewer people.

Borlaug's truism no longer holds, if it ever did. The modern organic movement began just before the green revolution, and it was similarly pioneered by researchers from top agricultural universities. These men and women recoiled from agricultural chemicals, many of which emerged from chemical-weapons labs of Germany and the United States during World Wars I and II. Lady Eve Balfour (1899–1990), the niece of British Prime Minister Lord Balfour, was a firebrand farmer, writer, and educator, and one of the first women to study agriculture at an English university. In 1939, at age 20, she began one of the earliest field trials comparing organic farming to chemical agriculture. Another pioneer of the organic movement was Professor William A. Albrecht (1888–1974), a prominent soil scientist at the University of Missouri who was president of the Soil Science Society of America. Albrecht was among the first to see a connection between poor soils and poor health. Austrian philosopher Rudolf Steiner (1861–1925) lent his own unique brand of science and mysticism to the movement, creating what he termed "biodynamic" agriculture. He advocated planting times based on the position of the moon and planets, which he felt influenced plant health.

The organic movement also evolved about the same time in Japan, when plant pathologist Masanobu Fukuoka (1913–2008) left his government research job to return to his family's farm on the island of Shikoku, where he spent the rest of his life developing a productive, small-scale organic farm system that didn't require weeding or tillage. His book *The One-Straw Revolution* was as much an essay on how to live as on how to grow vegetables and rice. "The ultimate goal of farming is not the growing of crops," Fukuoka wrote, "but the cultivation and perfection of human beings."

The most prominent and pugnacious of the early organic pioneers, however, was Sir Albert Howard (1873–1947). The son of a Shropshire farmer, Howard studied agriculture at Cambridge, where he trained as a mycologist—like Norman Borlaug, an expert in fungi and plant diseases. But Howard spent most of his career as an agricultural development officer for the Crown in India during the first quarter of the twentieth century. While breeding new varieties of cotton and wheat for the subcontinent at his field station in Indore, Howard found that improved seeds gave farmers about a 10 percent yield boost. But if they were planted in soils that were full of worms, fungi, composted manure, and humus—the dark carbon substance created by soil microbes when they break down crop residues and other organic matter—farmers could double their yields and grow plants so healthy they would stave off serious damage by insects, disease, and even drought. Sir Albert became the king of compost, creating it on a farm scale with green-manure crops and crop waste in what he called the "Indore Process."

Howard was as hardheaded as Borlaug and made many enemies. His 1931 book, *The Waste Products of Agriculture: Their Utilization as Humus*, landed like a cow pie on the desks of his fellow agricultural researchers. It questioned not only the merit of their work, but the need for them at all. Improved seeds, he argued, were far less important than improved soils. With fertile soils producing healthy crops, there was no need for chemical sprays or fertilizers or, by extension, chemists and entomologists to test them or companies to produce them. Howard's peers, in turn, excoriated him and ridiculed his work, provoking a response from him that might sound familiar to the anti-Monsanto protestors of today:

> The research workers on most other crops all over the Empire took a similar hostile view and were naturally supported and sustained in their opposition by vested interests like the manufacturers and distributors of artificial manures and poison sprays who were, of course, anxious to preserve and even expand a profitable business. It has been said that even the principle of gravitation would have had a hard row to hoe, had it in any manner stood in the way of the pursuit of profit and the operations of Big Business.

But Howard's verbal seeds fell on fertile ground. By the time he died in 1947, the Indore Process had spread to farms around the world, from England to South Africa, Canada to Chile, and India to Australia and New Zealand. The modern organic movement was born.

Howard's most influential convert, however, was not a farmer at all, but a young playwright from Manhattan's Lower East Side named Jerome Irving Cohen. After reading Howard's book *An Agricultural Testament*, which summarized his life's work, Cohen, who eventually changed his name to Rodale, an Americanization of his Polish mother's maiden name, wrote: "I was affected so profoundly that I could not rest until I purchased a farm. The reading of this great book showed me how simple the practice of the organic method could be."

Rodale bought land in the Lehigh Valley of Pennsylvania and proceeded to build not just a farm, but a publishing empire, producing books and magazines that focused on growing healthy food, exercise, and the prevention of disease. He published his first magazine title in 1942, and *Organic Farming and Gardening* (now *Organic Gardening*) became the bible of the organic movement in the United States. It was followed by *Prevention* magazine in 1950, which, along with several books he wrote advocating vigorous exercise, organic foods, and vitamin supplements, made him one of the nation's first health gurus. His belief that high meat and dairy consumption increased the risk of heart disease so angered the powerful meat and dairy industries and their allies in Congress that the Federal Trade Commission tried to prevent the sale of his books—a legal battle that lasted nearly two decades.

It was not without some irony that Rodale himself died of a heart attack while pretaping an appearance on the *Dick Cavett Show* in the early 1970s. But the surviving Rodales never stopped experimenting on their Pennsylvania farm. Jerome's son Bob Rodale and his wife Ardath eventually purchased a historic 333-acre tract of land near Kutztown, Pennsylvania, in 1972 and created the Rodale

Institute, a test farm where they conducted research on organic farming systems and provided advice to budding organic farmers. They wanted to fill the gaping void in such research at the USDA and land-grant universities, most of which, like my own alma mater, behaved like subsidiaries of the agrochemical industry.

Though Sir Albert Howard had eschewed plot-based research, Bob Rodale thought the only way to convince skeptical agronomy professors was to beat them on their own turf. So, in 1981 he worked with USDA researchers to set up a scientifically robust field trial to study the transition from conventional to organic fields and then compare organic and conventional yields. It took three years for the organic yields to rebound to levels comparable to the conventional fields, and from that point on Rodale's Farming Systems Trial evolved into an agricultural showdown: a scientific, side-by-side comparison of organic and conventional farming using the same crops on the same soils in the same climatic conditions. Now more than 30 years old, it is the longest-running study of its kind in the United States.

The Rodale Institute looks much like any other 200-year-old farm dotting the Pennsylvania countryside, with its 1790 clapboard farmhouse and two-story white barn trimmed in dark green. But the difference hits you square in the nose as soon as you step out of your car—a sweet mix of soil, compost, manure, fermenting silage, and verdant grass, with the faintest whiff of spring flowers. It smells like *soil*. When I visited, I couldn't help but think of my old soil science professor Joe Kleiss. To him, dirt was what you carved out from under your fingernails with your pocketknife. Soil was a sacred thing.

The old farmhouse had been turned into offices, and in one of those small rooms I sat down to talk with farm director Jeff Moyer, a middle-aged, barrel-chested farmer with a brown corduroy vest and an impressive silver Wyatt Earp mustache. Aside from managing the Rodale farm, Moyer is a former chairman of the National Organic Standards Board. His roots run deep in the Pennsylvania soils. One of his ancestors nine generations back was born in the old farmhouse, and he was raised on a small farm nearby. Though he studied forestry in college, he became interested in the back-to-the-land movement and got a job as a field hand at Rodale in 1976. He's been there ever since. But managing the most visible organic farm in the nation has not been easy.

"When I became farm director, Bob Rodale's challenge to me was this: you have to make money every year. You have to do research as well. And at the same time you have to improve the soil. If all you do is produce food and don't improve the resource, you're fired. If you just improve the resource but don't produce food, you're fired. You have to do both."

"It's a wonderful philosophy," I said. "But with the population expected to hit 9 billion by 2050, and experts saying we need to double our food production by then, can this system feed the world?"

"If we talk about feeding the world for the next 50 years, then the conventional system is fine," Moyer says. "There are ample resources to destroy for the next 50 years. But what about 5,000 years from now? What about 15,000 years from now? Anyone who tells you we'll still be using Roundup in 5,000 years is crazy." The more immediate problem, he says, is our system of subsidies, which pays growers to produce vast amounts of corn (which he calls "the yellow stuff") instead of a mosaic of crops. "The answer to the argument that organic agriculture will never feed the world is that we've never tried. And think of the billions we've spent propping up the other way."

We took a walk and headed over to the Farming Systems Trial, which lies on a field at the edge of the property, where the pesticide and fertilizer applications on the conventional plots won't jeopardize the organic certification of the rest of the farm. The Rodale Institute supports a 200-member CSA program. Members pay $700 at the beginning of the year and receive a box of fresh organic vegetables every week for 22 weeks during the harvest season. The farm also provides boxes of vegetables to food stamp outlets in nearby cities for $15 per week, where Rodale staff and volunteers teach cooking classes and provide farm tours for city kids.

Rita Seidel, the Farming Systems Trial manager, joined Moyer and me at the trial plots. Seidel earned her degree in Germany, the country with perhaps the world's longest tradition of organic agriculture. We walked to a rather dilapidated viewing stand from which we could see the 12-acre test field neatly divided into 72 plots, each 20 feet wide by 300 feet long, wide enough to plant eight rows of corn at the typical 30-inch spacing. The conventional plots are planted in a corn-soy rotation (with wheat added in 2004), using best management practices published by Penn State, Pennsylvania's preeminent agricultural university. The organic manure system—designed to replicate a mixed farming operation with crops and livestock—grows corn, soy, hay, and alfalfa, with cover crops of rye and corn cut for silage in some years.

The organic legume plots were designed to mimic a "cash-grain" rotation of corn, soybeans, and wheat, with rye and hairy vetch (a nitrogen-producing legume) planted as green-manure crops. The hairy vetch immediately stood out in the plots—a thick swath of knee-high plants whose tiny purple flowers were sprinkled through the fields. It was hard to tell whether the plots looked more like a miniature jungle or a shaggy lawn in desperate need of mowing. But the vetch is a nitrogen-making machine. The farmer's rule of thumb is that every pound of applied nitrogen per acre produces a bushel of corn. Vetch can produce 300 pounds of nitrogen per acre—almost too much, said Seidel. In years with heavy rainfall, both the manured fields and those with vetch leach excess nitrogen into the water table, though leaching occurs much less often from the organic fields than from the chemically fertilized fields.

Nitrogen isn't the only thing that green-manure crops provide. The lush stems and leaves of the vetch eventually break down into organic matter in the soil. For soil scientists, increasing organic matter means more porosity, more water-holding capacity, more gaseous exchange, more cracks and crevices for roots to grow. It means more fungi, earthworms, nematodes, grubs, bacteria, arachnids, insects, and all the other wondrous soil organisms that help make soil fertile. More organic matter in soil is thus a primary goal of organic farmers—and one that is shared by climate scientists, who have estimated that the world's soil contains roughly 2,500 pentagrams of carbon.

That's a pretty mind-blowing number. One pentagram equals a billion metric tons. Thus, the world's soils contain more than three times the carbon currently floating around in the atmosphere and four times the amount currently tied up in forests and plants. Global agriculture, with its sod-busting plows has released 50–70 percent of the original carbon stored in soils into the atmosphere. And therein lies the true, potentially climate-saving aspect of organic agriculture. If farmers could increase soil organic matter (which is almost 60 percent carbon by weight) on the world's farmland, they could sequester as much as a third of annual global carbon emissions and help grow healthier, more drought-resistant crops in the bargain.

Rodale's trial has produced other dramatic results. Like Sir Albert Howard's experimental farm in Indore, the organic plots suffered significant yield losses as they were weaned off chemicals. Organic corn yields averaged almost 30 percent lower than conventional corn from 1981 to 1985, while organic soybeans—which fix their own nitrogen—actually yielded significantly higher than conventional beans during the first five-year rotation, but their yields were hampered by heavy weed pressure and the shorter-season varieties that Rodale used during the second five years. Small grains like wheat and barley were not grown in the conventional plots, but from the beginning of the trial the organic plots matched the yields of wheat, oats, barley, hay, and corn silage that conventional farms in the county achieved.

After a few years, however, the quality of the soil in the organic systems began to improve dramatically. Researchers found better soil structure (what agronomists call "tilth" and what my children might describe as "crumbliness"), increased organic matter (2.5 percent versus 2 percent), and higher biological activity of the various soil inhabitants, from microbes to fungi to earthworms, which keep the soil fluffy and aerated like a big living sponge. After the transition period, the organic plots began producing robust crops that matched the yields of the conventional plots despite far higher weed pressure—a marketable trait in a world of herbicide-resistant weeds. The fields are not irrigated, and during drought years, corn yields in the organic plots actually were 31 percent *higher* than yields of conventional corn. Not surprisingly, water volumes contained in the organic soils were 15–20 percent higher than those in the conventional plots.

More important from an environmental perspective is what the organic fields did not yield. Nitrate leaching occurred far more often in the conventional fields, frequently exceeding the EPA limit of 10 parts per million for safe drinking water. Leachate from the conventional fields also contained the grass herbicide atrazine—at times exceeding EPA levels for drinking water. The organic systems used 45 percent less energy overall than the conventional systems, and emitted 40 percent less of greenhouse gases per pound of crop grown. Other studies outside of Rodale have found that the external costs of an organic farm—such as soil erosion, the contamination of watersheds with pesticides or fertilizers, and adverse effects on birds and other biodiversity—were just one-third those of conventional farms.

The economic benefits were also substantial. Rodale's organic plots earned almost triple the profit of the conventional plots ($558 per acre per year for organic, versus $190 per acre per year for conventional), thanks to the high premiums paid for organic crops. But even without the organic price bonus, the organic returns mirrored those of conventional farms, staying competitive because of the lower input costs.

The downside was that organic farming took more work (27 percent more hours of labor per acre during the first 15 years) and more-intensive management. And the organic plots had to rotate crops for pest control and soil fertility. They couldn't grow corn after soybeans year after year on the same fields as some conventional farmers do. They also experienced a significant yield drag during the transition. After that, however, organic corn yields on the Rodale farm have averaged about 130 bushels per acre—right at the county average for conventional farms in the rest of Berks County, Pennsylvania.

It's hard for farmers, as well as agricultural researchers, not to be seduced by high yields. In 2008, just down the road from Rodale, a Berks County conventional farmer named David Wolfskill won the National Corn Growers Association yield contest for no-till dryland corn, harvesting 297 bushels per acre, more than double the county average.

"If the weather cooperates, conventional corn is like a thoroughbred," says Rodale farm manager Jeff Moyer. "When the track is perfect, that thing can sure run. But when is the track ever perfect, especially with climate change? That's why conventional agriculture is going out the door. So we're trying to take the racehorse and turn it into the draft horse. It's not going to win the Kentucky Derby, but we're going to grow food and make money every year."

Rodale, however, is a high-achieving farm. Other studies in the United States and Europe, including large meta-analyses have found that organic crops typically yield anywhere between 6 and 20 percent less than conventionally grown crops, with the most recent study in 2012 showing a 13 percent difference when best organic practices were used. But the results varied highly depending on water availability and the crop measured. Organic farmers typically do far better on rain-fed fields, and the yields on crops like wheat and tomatoes were virtually identical.

In the developed world—especially in the breadbaskets of North America and Europe, where irrigation and intensive fertilizer and pesticide use are widespread—conventional farming wins the yield contest hands down. In the developing world, however, where soils are poor, yields are low, and agricultural chemicals are either unattainable or prohibitively expensive, the yield gap between the two systems vanishes. In 2005, Rachel Hine and Jules Pretty at the University of Essex analyzed nearly 300 sustainable agricultural projects covering 37 million hectares in 57 developing nations. The study found that when farmers adopted organic or sustainable agricultural practices integrating a variety of crops or systems, average crop yields increased nearly 80 percent, with yields in Africa more than doubling. A similar study that tracked a thousand Indian farmers for seven years in the arid Maikal district of central India found that yields of soy, cotton, wheat, and chili peppers were as much as 20 percent higher on organic farms than on nearby conventional farms, thanks to composting and increased organic matter that boosted the water-holding capacity of the soils.

The number of farmers adopting organic methods—or even quasi-organic methods—continues to grow around the world, driven by both the environmental and the economic benefits. The case of SRI, short for the "system of rice intensification," is a classic example. SRI was developed in the 1980s by a French Jesuit missionary named Henri de Laulanié, who spent the last 34 years of his life living and working with farmers in rural Madagascar. Before attending divinity school, de Laulanié earned his undergraduate degree from the top agricultural university in France. The Jesuits sent him to the impoverished island nation in the 1960s, where average yields were about 2 tons per hectare, even with government-subsidized fertilizer.

Like Sir Albert Howard, who studied peasant agriculture in India, Father de Laulanié spent a lot of time observing traditional Malagasy rice farmers. After more than a decade of trial, error, and fortunate accident, he developed a method that ran totally counter to the green revolution. Instead of transplanting large seedlings in clumps close together, he planted tiny seedlings, 10–12 days old, 25 centimeters apart, so that each tender shoot would have nearly a square foot of soil to itself. And instead of keeping the paddy flooded to suppress weeds—an agricultural technique passed down for millennia—he merely kept the soil moist and used a handheld rotary hoe (anyone remember the multipronged Garden Weasel?) to weed and aerate the soil. This method had the added benefit of reducing methane emissions from rice paddies. (Global rice fields generate 13 percent of global methane emissions, which have 20 times more impact on climate change than carbon dioxide does.) When the government of Madagascar phased out fertilizer subsidies in the late 1980s, Father de Laulanié shifted to compost, creating a totally organic production system.

De Laulanié's results were counterintuitive as well. Even though his method used half of the water of conventional rice paddies, 10 percent of the seeds, and little if any fertilizer, the SRI farmers quadrupled their yields to 8 tons per hectare, with some fields on the best soils yielding an uncanny 16 tons per hectare. In the early 1990s, de Laulanié and his friends founded an agricultural school for boys, as well as a small nonprofit group to promote the new system. He named the group Tefy Saina—which in Malagasy means "to improve the mind."

SRI turned out to be as controversial as the Indore method nearly a century before. With the exception of the Cornell International Institute for Food, Agriculture and Development, which has worked with SRI farmers and Tefy Saina to promote the technique since the mid-1990s, the bulk of the international agricultural community has shunned the method.

Dr. Amrik Singh, an enthusiastic young agronomist with the Punjab Department of Agriculture, is an exception. Even though Punjab, an epicenter of the green revolution, is now beset with water woes, Singh had to get special permission from his superiors, along with grant money from the World Wildlife Fund, to introduce SRI to 20 small farmers in Punjab. With all the potential water-saving benefits of the method, I asked him why Punjab Agricultural University or the International Rice Research Institute was not actively promoting SRI to farmers.

"They haven't invented this thing," Singh said. "They have published none of the literature. This benefits only the farmers [who] use a rope to knock insect pests into the water where they are eaten by beneficial insects like the damselfly and dragonfly instead of spraying pesticides. The [fertilizer and pesticide] traders have no interest in this either. They know that if this takes hold, they will sell less of their products and make less money. But if we apply SRI to 2.6 million hectares of rice in the Punjab, we could save 50 percent of the water and electricity, and more than half of the seeds."

It wasn't easy to get Punjabi farmers to try the "Madagascar method." We drove to the farm of Keshav Bahl, one of the more prosperous farmers in the region who was planting SRI rice for his second season. Bahl, a round, soft-spoken man, had been farming since 1960 and had actually met Norman Borlaug once at Punjab Agricultural University. Like many of the wealthier farmers who initially profited from the green revolution, Bahl has watched those profits wither as the cost of inputs like diesel and fertilizer has risen. He still longs for a repeat of those glory days. "When Borlaug brought in the new Mexican varieties, there was a three-fold increase in cereal production," Bahl says. "It was a great relief. But now the green revolution is totally exhausted. Farmers are not happy. Costs have gone up, but returns are much less. When a farmer spends 1,000 rupees per hectare, he must get 1,500 rupees in return. Now we only get 1,050."

Bahl, who is a seed producer and has access to capital and plenty of green-revolution inputs on his 40-acre farm, was already producing high yields from his conventional fields. Yet with SRI he boosted those yields and cut his input costs, applying fertilizer at only a third of the recommended rate. When Singh and I walked with Bahl out into his SRI rice paddy, which was nearly as dense as a lawn and 2 feet high, Bahl seemed pleased by the results. The SRI plants were thicker, taller, and stronger than his conventional rice plants, with a significantly higher number of the branches known as "tillers." Each tiller supported heavy, rice-laden panicles. Bahl pulled a plant from each field and held them side by side. The SRI plant dwarfed the conventional rice plant. Those deep, thick roots, Singh explained, helped the rice plant to mine water and nutrients from the soil. In fact, Cornell researchers measured the force required to pull an SRI rice plant out of the ground and found that it was five to six times higher than for conventional plants. The previous season, Bahl's conventional fields had yielded 20 quintals (there are 100 kilograms per quintal) per acre, while the SRI fields had yielded 28 quintals. That amounts to roughly 6.9 tons per hectare versus 4.9 tons per hectare—a 40 percent increase.

IF BAHL WAS optimistic about the future of SRI plants, Makman Singh, a 40-year-old farmer who tends 5 acres outside the town of Dhariwal, could barely contain his excitement. As we approached, the farmer grasped Dr. Amrik Singh's hand and thanked him profusely for introducing him to SRI, and then offered us fruit juice, biscuits, even whiskey as he told us his story.

"My father was really against it," Makman Singh said. "He doesn't want to do anything new." Even so, Makman had planted about an acre of SRI rice as an experiment, and when we walked to the field, the reason behind his enthusiasm became apparent. During my tenure as a field editor for *Rice Journal* or through my travels in Indonesia, I had never seen a lovelier paddy of rice—thick and lush with every panicle full of seeds. Amrik Singh had previously sampled the field and estimated its yield at nearly 10 tons per hectare, more than double Makman's average yield for conventional rice. The next season, Makman Singh planned to plant his entire farm using SRI. "What does your father say now when he sees this field?" I asked the farmer. "He does not say anything now." Makman said, with a grin as big as the sky.

SRI remains controversial among traditional rice researchers. Thomas Sinclair, a prominent plant physiologist with the USDA's Agricultural Research Service, dismissed SRI in a blistering 2004 article in the IRRI publication *Rice Today*: "Discussion of the system of rice intensification (SRI) is unfortunate because it implies SRI merits serious consideration. SRI does not deserve such attention," Sinclair wrote. In the Philippines, IRRI researchers could get no better than 3 tons per hectare with SRI, less than the Philippine average yield of 4 tons per hectare. Critics at IRRI and elsewhere point out that many subsistence farmers quickly abandon the method because it requires more management and labor, and thus it isn't sustainable over the long haul.

The most formidable argument in favor of the Madagascar method was made not by researchers, but by a quiet young farmer in Bihar, the most impoverished state in India. Shri Sumant Kumar typically harvested 4–5 tons of rice per hectare on his fields near the Sakri River in the Nalanda district. In 2012, he and four friends experimented with SRI, which was being promoted in their area by a local nonprofit group. Each farmer planted 1 acre of SRI rice, as well as 5–7 acres of conventional rice, on their 2- to 3-hectare farms. Kumar and his friends actually used a blend of SRI and green-revolution techniques on their fields, incorporating a relatively low rate of chemical fertilizer in both conventional and SRI fields, along with green-manure crops of sesbania that were plowed under before planting. The farmers used mixed rotations prior to their rice plantings, most of which included legumes. They then planted conventional fields with large seedlings at the normal, high-density rate and controlled weeds with herbicides and flood irrigation. All the fields were planted with high-yielding varieties of rice.

The rains that season were very good, and Kumar had high hopes of a bumper crop. Instead, he harvested an almost unbelievable 22.4 tons of rice per hectare from his SRI field, measured on his village's old scale. This yield shattered the previous world record of 19.4 tons, held by none other than Dr. Yuan Longping, a Chinese agricultural scientist long known as the "father of rice." It also far exceeded any crop ever produced by the rice researchers at IRRI. And Kumar was not alone in his good fortune. His four friends each topped 17 tons per hectare on their SRI fields. When the word got out, many accused Kumar of cheating, until the state director of agriculture—also a rice farmer—came to the village with his technical staff and personally verified the yields.

Many agricultural scientists still brushed off Kumar's rice record and SRI as a fluke—until another of Kumar's friends broke the world yield record for potatoes using SRI a few months later, and then another nearby SRI farmer broke India's national record for wheat yields. Yield increases have been recorded for SRI sugarcane, yams, tomatoes, garlic, and eggplant as well. The method has been credited with increasing the entire region's yields by 45 percent.

Moreover, the SRI plots cost significantly less to produce, saving farmers 2,900 rupees per hectare, or 32 percent less than the cost of conventional rice, netting more profits for the farmers. The most revelatory thing to Norman Uphoff, the former director of the Cornell International Institute for Food, Agriculture and Development, wasn't that the yields broke records, but that the same farmers who achieved those records averaged only 7 tons per hectare on their conventional rice fields—same farmers, same soils. Uphoff credits the difference to the increased symbiotic relationship between SRI plants and a host of beneficial fungi in the soil that researchers have recently discovered can promote root growth and even turn on and off certain genes in the rice plant.

"The agricultural establishment said SRI was like going back to the dibble stick," says Uphoff, referring to the pointed planting stick used by farmers for millennia. "Nothing could be further from the truth. We're talking about symbiosis, from the uptake of micronutrients to enzymes that regulate metabolism. This is postmodern agriculture."

Such results seem to be winning over the critics. Sun Lu Ping, one of the top rice experts in China, has supported SRI for a decade now, as has India's legendary rice researcher M. S. Swaminathan, who was Borlaug's close friend and the father of the green revolution in India. Ohio State University's Rattan Lal, one of the top soil scientists in the United States, has also endorsed SRI. Departments of agriculture in India, Vietnam, Cambodia, Indonesia, Bangladesh, and China are all promoting the method, which is now practiced by more than 4 million farmers around the world.

ANOTHER CRITICISM OF organic agriculture that is now falling by the wayside is that it doesn't lend itself to large operations. While reporting on Brazil's sugarcane ethanol industry in 2005, I stopped by a sugar mill owned by the family of Leontino Balbo

Jr., the most enthusiastic organic farmer I've ever met. He began trying to grow what he calls "green cane" in 1986, and by 1997 he had certified 15,000 hectares of organic sugarcane, creating one of the largest organic farms in the world. Some of that cane was also certified biodynamic—raised with the Austrian method established by Rudolf Steiner that regulates planting dates according to a celestial calendar, among other requirements. Biodynamic sugarcane is sold in Germany at a premium. Balbo's sugar mill was producing 80,000 metric tons of organic sugar each year—at the time that was 38 percent of the world's supply—along with 60,000 cubic meters of organic ethanol. Under the Native brand, Balbo now sells to 55 countries and is a major supplier to Dannon yogurt and Starbucks.

Like many other organic farmers, Balbo had to overcome numerous obstacles. First he had to find a biological control for the sugarcane borer, the crop's most destructive pest. He eventually found a parasitic wasp that was the borer's natural enemy, and each year his workers release millions of wasps into the fields. Next he developed a new type of sugarcane harvester. Most cane fields are burned before harvest, which makes hand cutting easier but creates enormous air pollution and kills critical microorganisms that live in the top layer of the soil. So Balbo worked with an equipment manufacturer to develop a machine that would cut green cane. But perhaps the hardest hurdle to overcome was the skepticism of the Balbo board of directors—his father and uncles who had been farming the land since 1946.

"The farmers in São Paulo are very conservative," Balbo said. "They are old Italians, Portuguese, and Germans. Very hard to change. For 12 years we were defeated. But now our yields are 20 percent higher than the São Paulo average, with 10 percent less production costs." For the previous seven years, Balbo said, they had grown 105 tons of cane per hectare, compared to the 85-ton-per-hectare average for São Paulo. Moreover, most conventional cane—a perennial crop—has to be reestablished every five years. But the organic fields typically last seven. During a drought year, he still harvested 93 tons per hectare, while the state average fell to 68 tons. The most powerful argument to his elders, however, was the profit. At the time of our interview, Balbo was selling his conventional sugar for $250 per ton, organic sugar for $550 per ton, and biodynamic sugar for a whopping $950 per ton.

We hopped into his Land Rover and blasted down the field paths at speeds only a Brazilian would find amusing, passing through a rolling sea of cane 3 meters high, with Balbo talking nonstop about the benefits of organic production. It was good not only for the family finances, but for the farm's wildlife as well. When biologists surveyed the farm, they found rich biodiversity, he said, including nine species of ladybugs per hectare (two types previously unknown to scientists) and 100–200 million earthworms per hectare. The farm held 2,300 capybaras, the giant rodents of Brazil, which have attracted back pythons, including one 11-meter monster that they photographed and released (another reason not to harvest the cane by hand). They also found a glass snake, actually a lizard, long thought to be extinct in the state of São Paulo, as well as some 44 amphibians, nearly as many as the 48 species of amphibians found in the largest protected reserve in Brazil.

Even small farmers around the globe are finding that if they simply incorporate a few of the best organic or agroecological practices, they can reap both financial and environmental benefits— just as Shri Sumant Kumar and his friends in Bihar discovered. A recent study by FAO researchers Amir Kassam and Hugh Brammer found that the use of Conservation Agriculture (which minimizes soil disturbance, maintains a continuous top layer of organic mulch and cover crops, and utilizes a mix of crop species, including annual crops, trees, shrubs, and pastures in rotation) and the system of rice intensification has reduced the costs of production, increased yields, and provided important environmental benefits in some of the hungriest parts of the planet. When the techniques were introduced in the Karatu District of Tanzania in 2005, average maize yields leapt from 1 ton per hectare to 6 using only farmyard manure for fertilizer.

Results were similar for SRI. In a survey of more than 700,000 SRI rice farmers in Vietnam, only 20 percent were using all the practices recommended for the technique. Yet even those using semi-SRI saw 10–14 percent yield increases with a 30 percent savings in water and a 32 percent reduction in production costs.

"In the coming decades, both [Conservation Agriculture] and SRI appear to offer the best hope of increasing food production rapidly, at low cost and without adverse environmental consequences in developing countries where human populations are increasing most rapidly," the authors concluded.

Just imagine what the world could do if we took that lesson even further and transformed the powerful swords of genetic engineering, public research, and agricultural subsidies into agroecological plowshares.

Before leaving the Rodale Institute, I dropped by the office of the new director, Coach Mark Smallwood. Smallwood, a big man with a big shiny pate, comes by the "coach" moniker honestly, having spent much of his career as a basketball coach at Cooper Union, as well as a few years as an assistant coach under Pat Riley with the New York Knicks. But his green roots run deep as well. He's a lifelong organic farmer, beekeeper, oxen driver, and goat herder, and he was a local "forager" for Whole Foods, sourcing organic produce from local farmers for Whole Foods grocery stores throughout much of the Mid-Atlantic region. Two prominent signs were hanging on his office wall: one read "Create a Massive Awakening"; the other, simply "CHANGE" in huge block letters. I asked him how he plans to foment such change in a world where the green revolution, and the large corporations and institutions it created, hold such powerful sway. Smallwood responded as if giving a halftime pep talk to the Knicks:

"Right now we've got 4 percent of the food and beverage sales in the US. It's 96 to 4 and we're getting our butts kicked, but at least we're in the game. Well, a few months ago Rensselaer Polytechnic Institute did a study that showed that once a population gets to 10 percent agreement on anything, that's the tipping point for it to go majority. Not 51 percent. How many years will it take us to get to 10 percent of food and beverage sales? I think we'll be at 51 percent by 2027. It's consumer power. We've got four times the job creation than the national average, 94 percent of organic operations nationwide are maintaining or expanding, and 78 percent of families are buying some organic product in all 50 states. We've got 4.6 million acres and growing. It's the exciting place to be in the world of agriculture!"

I thanked Smallwood for his time, and as he walked me to the door of his farmhouse office he told me a story that he swore was true and emblematic of the great paradigm shifts he sees under way in the food system at large:

"I was milking my goats in my barn one day when a man walked in hoping to buy some honey. 'Sure,' I said. 'Just wait until I'm done here. In the meantime, do you know what happens when equilibriums clash?'

'I guess not,' the man said.

'Shift happens,' I replied. 'Paradigms are unconscious road maps. Now, would you like some goat's milk while you wait?'

'Uh, no thanks.'

'How about a Coke?'

'Sure.'

'So you want carbonated sugar water, which has absolutely no nutritional value, while the raw goat's milk I want to give you is one of the finest foods on the planet?'

'Okay, okay,' the guy said. 'I guess I'll try it.' So I handed him a Dixie cup full of goat's milk and he took a sip. Then he said . . . and I kid you not, 'Mmmm. That's some good shift.'"

Alice Sumphi, a farmer in the Soils, Food, and Healthy Communities project in northern Malawi, proudly shows visitors her first tomato crop.

CHAPTER 13

The Malawi Miracle

Let us pick any Malawi road, you are likely to find a woman at mid-day, possibly with a baby on her back, and that woman would have had no breakfast, perhaps even no water to drink, and you in the West are saying, she does not need subsidy.—*Bingu wa Mutharika, 2009*

The road to Zomba is fast and good, a black bitumen snake lying across the rolling red soil of the Lilongwe Plain. The plain sits atop a jumbled plateau creased with ridges and punctuated with massifs that follow the eastern branch of Africa's Great Rift Valley, forming the foundation of the small nation of Malawi.

It was October, the end of the dry season. Jacaranda and flame trees added splashes of purple and crimson on an otherwise burnt-umber canvas that stretched toward dusty green mountains in the distance. There were real flames as well. October is the burning season here. Plumes of black smoke rose in the distance as local farmers torched weeds and crop residues from their tiny plots of land. The small fields beside the road were tilled into deep, red ridges, waiting for the seasonal rains to soak and soften them before they received precious seeds. The ridges testified to millions of hours of toil with a short-handled hoe, the ubiquitous tool of the small-plot farmers who produce the vast majority of Africa's food. Yet this same agricultural system produces chronic shortfalls and grinding poverty that seems to have no end.

The road to Zomba is also a microcosm of the world food crisis. Africa is the continent where *Homo sapiens* was born, and with its worn-out soils, vanishing forests, erratic rains, and booming population, it could foretell our species' future. The vast majority of the world's 795 million malnourished people live in rural areas of South Asia and sub-Saharan Africa much like Malawi, where half the children under five are stunted from poor nutrition. Hunger-related illnesses kill an estimated 4 million Africans every year. Most are children. Each year hunger kills more people around the globe than AIDS, malaria, and tuberculosis combined.

Despite billions spent by western aid agencies to promote it, the green revolution never really took hold here. The reasons were both physical and political. When Norman Borlaug turned his attention to Africa in the 1980s, he encountered stiff resistance from US and European environmental groups that strongly opposed exporting chemical monocultures to the continent, home to spectacular biodiversity and the second-largest virgin rain forest on the planet. Such groups convinced the Ford and Rockefeller Foundations and the World Bank to shift funding from agricultural projects to environmental and health causes, while European Green Parties pressured their governments to cut funding for fertilizer shipments there.

Borlaug, always the fighter, blasted his critics. "Some of the environmental lobbyists of the Western nations are the salt of the earth, but many of them are elitists," Borlaug told the *Atlantic Monthly* in 1997. "They've never experienced the physical sensation of hunger. . . . If they lived just one month amid the misery of the developing world, as I have for fifty years, they'd be crying out for tractors and fertilizer and irrigation canals and be outraged that fashionable elitists back home were trying to deny them these things."

In his 70s Borlaug was pulled from retirement by Japanese industrialist Ryoichi Sasakawa to head the Sasakawa Africa Association in 1986, which eventually joined forces with President Jimmy Carter's Global 2000 initiative to get the green revolution rolling in Africa. Borlaug remained president of the initiative until his death in 2009. Using the same strategy that he had employed in the Punjab, the group planted thousands of large demonstration plots in 14 African nations, showing farmers and extension agents what good seeds and fertilizer could do. In Ethiopia the plots doubled the average yields of wheat and teff and more than tripled the average yields of maize.

But as Borlaug knew all too well, a green revolution needed more than test plots. Two centuries earlier Adam Smith blamed the lack of good transportation routes for keeping Africa in a "barbarous and uncivilized state." The problem lingers today. To follow in the footsteps of Mexico and Punjab, African nations needed ports to bring in subsidized seeds and fertilizer, and good roads to get them into the interior and get the crops out to market. They needed subsidized irrigation schemes to protect farmers from drought, which required huge sums for dams and canals for surface irrigation, or expensive pumps and power (electricity or diesel) to tap groundwater. They needed farm credit so that smallholding farmers could purchase expensive fertilizer and seeds that cost more than their current income. Most African farmers don't even have a proper title to the plots they till, leaving them with little, if any collateral.

Most important, a green revolution needed the full support of a functioning, noncorrupt government that is not at war with its citizens or its neighbors—a commodity as scarce as fertilizer in postcolonial Africa. In 1990 I covered a speech at the United Nations by Mobutu Sésé Seko, at the time "president for life" of the former nation of Zaire. He was renowned for his bespoke suits, leopard-skin hat, and mind-bending thievery. I later learned from a Peace Corps volunteer in the country that most of the food is grown in one region, and most of the people live in another. No good road connects the two, although one had been proposed, designed, and funded several times. Mobutu, who allegedly received $150 million in CIA payments to be our Cold War ally, as well as more than a billion dollars in US development aid, never built the road. But he did buy a château in Belgium, a castle in Spain, and a "Versailles of the Jungle" on the banks of the Congo. His personal fortune at his death in 1997 has been estimated at more than 6 billion pounds sterling—enough to pay off his country's entire foreign debt.

And Mobutu was not alone. Tanzania received $2 billion in aid for road building over the last two decades without any noticeable improvement in its road network, while an endless river of oil, minerals, precious metals, and timber flowed out of the resource-rich continent with almost nothing to show for it at home. Though such numbers are almost impossible to track, one recent study estimated the total illegal capital flight from Africa between 1970 and 2010 at almost $700 billion. That's more than double Africa's crushing foreign debt in 2009 of $300 billion. If such graft had been put to good use, African nations could be debt-free and have $400 billion for education, vital infrastructure, and agricultural projects.

Instead, the money just vanished, mostly into the foreign bank accounts of political elites and their extended families. Sub-Saharan Africa is the only region on the planet where food production per capita actually *decreased* during the green revolution. Farmers in Asia now use almost 100 kilograms of inorganic fertilizer per hectare, while in sub-Saharan Africa the figure hasn't budged from an average of 9 kilograms per hectare for decades. Since 1960, average yields across the African continent have remained at about 1 ton of grain per hectare—roughly the same amount British farmers were growing under the yoke of the Roman Empire. The continent now has an annual 10-million-ton food deficit, and it is home to more than a quarter of the hungriest people on Earth. If population, income, and urbanization continue to grow at the current projected rates, Africa will need to *more than triple* its food availability by 2050 for everyone to be adequately fed.

"Sub-Saharan Africa is where the action is," says soil scientist and World Food Prize recipient Pedro Sanchez of Columbia University, who was cochair of the UN Millennium Project's Hunger Task Force. "And it's where these issues need to be resolved."

SANCHEZ PROUDLY DESCRIBES himself as "a foot soldier in the green revolution," and in 2005 he seized the chance to spread the revolution to Malawi. That year the rains failed again, and 5 million Malawians required food aid from abroad. Newly elected president Bingu wa Mutharika sought the advice of Sanchez and his colleague, renowned economist Jeffrey Sachs. At the time, Sachs and Sanchez were launching the Millennium Villages Project, a high-profile campaign to show, among other things, the benefits of bringing green-revolution inputs to Africa. Two of the experimental villages are in Malawi.

"I did not get elected president to rule over beggars," said Sanchez, recalling Mutharika's words at the height of the food crisis in 2008. "So we told him to subsidize nitrogen and hybrid seeds."

Fertilizer subsidies weren't new to Malawi. The nation's former dictator Hastings Kamuzu Banda had subsidized agricultural inputs to some extent throughout his thirty-year reign. Like many poor developing countries, Malawi's debt skyrocketed in the 1990s. The World Bank and International Monetary Fund, the primary lender to poor governments, instituted structural adjustment programs that forced borrowing nations to cut agricultural spending. Though aimed at trimming national debts, the policy, as Mutharika noted, was deeply hypocritical. Europe, the United States, and Japan, primary funders of the World Bank and IMF, have some of the most subsidized farmers on the planet.

When Mutharika asked the World Bank and foreign donors to help finance fertilizer for his farmers, they turned him down. Bringing back the subsidies had been a main platform of his 2004 campaign, and Mutharika was furious. He spent $58 million from his own government coffers to provide more than a million poor farm families with two 50-kilogram bags of fertilizer and a 3-kilogram bag of hybrid maize seeds. They paid about a third of the market price for them.

What happened next—whether due to Mutharika's brilliance or his uncanny good fortune—has been dubbed the "Malawi Miracle." Normal rains returned to the land during the next two growing seasons, and with good soil moisture, high-yielding hybrid seeds, and a modest amount of fertilizer in the ground, Malawi's farmers reaped back-to-back bumper crops of maize. The government estimated the 2007 harvest at 3.44 million metric tons, the greatest corn crop Malawi had ever grown. "They went from a 44 percent deficit to an 18 percent surplus, doubling their production," said Columbia's Sanchez. "The next year they had a 53 percent surplus and exported maize to Zimbabwe. It was a dramatic change."

Critics of the program said Malawi just got lucky with the return of good rains. But Sanchez disagreed. "We estimated that the effect of the rains contributed about 30 percent of the increase, while the nitrogen and good seeds contributed the rest," he said. "You make your own luck."

The Malawi Miracle had an immediate impact. The *New York Times* wrote a story lauding Mutharika and vilifying the World Bank that was headlined ENDING FAMINE, SIMPLY BY IGNORING THE EXPERTS. In 2006 and 2007, the Rockefeller Foundation and the Bill & Melinda Gates Foundation launched the Alliance for a Green Revolution in Africa, headed by former UN secretary-general Kofi Annan, pledging hundreds of millions of dollars to bring plant-breeding programs to African universities and to get more fertilizer into farmers' fields. Jeffrey Sachs and Pedro Sanchez lobbied the World Bank to reverse their policies against agricultural subsidies, which they soon did. "Smart subsidies" became the new development buzzword. Mutharika was feted and awarded prizes all over the globe. Other African leaders jumped on the bandwagon, and as of 2012, nine more nations had instituted large-scale fertilizer subsidy programs. Total cost: roughly $2 billion each year.

THE MILLENNIUM VILLAGES PROJECT is by far the most ambitious development project to land in Malawi. Sachs was a major architect of the eight Millennium Development Goals that were adopted by all members of the United Nations in 2000, aimed at halving poverty, hunger, and disease by 2015. Since many of these goals are interdependent, Sachs believed they should all be tackled together at the same time. To prove his "big push" theory, Sachs and colleagues sent huge amounts of aid and technical support to 14 impoverished village clusters in 10 African countries to help them meet all Millennium Development Goals in just five years. A key goal is improving agricultural production. Some economists have estimated that as much as 54 percent of the poverty reduction in all the developing world between 1960 and 1990 came directly from green-revolution yield increases in Asia and Latin America. Another 29 percent was indirectly attributed to agriculture, since the increasing yields enabled some laborers to move into more productive sectors of the economy.

Sachs has argued that the green revolution laid the foundation for rapid economic growth in China and India and it could do the same for Africa. But big pushes cost money—to the tune of an estimated $27.1 million per village in 2011. The staggering cost of the project, not to mention the glaring inequality created between the Millennium villages and those surrounding them, has roiled other development experts. Many doubt whether such an investment is sustainable, scalable, or even effective in the long run.

The project's ample resources were obvious when I met Phelire Nkhoma, the agricultural extension officer for Mwandama, one of Malawi's two Millennium villages. At her office in Zomba, located in the largest bank building in the former colonial capital, every desk had a working computer, and air conditioners hummed against the October heat. When we walked outside to start our tour, Nkhoma ushered me into a new, gleaming-white Toyota Hi-Lux diesel pickup—a $50,000 vehicle that is the limo de rigueur of the development world—with blue UNDP emblems on it. The UN agency is a partner in the effort.

Nkhoma herself was a small, humble, soft-spoken woman in a brown cotton print dress, with an enormous laptop and a navy baseball cap that read "JESUS IS LOVE." As we rode the bumpy back roads to the village, she explained the basics of the program to me. The Mwandama Millennium village is actually a cluster of seven small villages that are home to some 35,000 people. Sachs's big push covers many of the basic needs that are lacking in much of rural Africa. Villagers get hybrid seeds and fertilizers for free as long as they donate three bags of corn at harvest to a school meal program. They get bed nets and antimalarial drugs. They get a clinic staffed with trained health workers, as well as a large, well-built warehouse to store their seeds, fertilizers, and harvested maize. Safe drinking-water wells were built within a kilometer of each house. New primary schools, improved roads, electric power, and the Internet were also on the way to Mwandama, as well as to the Madonna village farther north.

"*The* Madonna?" I asked. "Yes," Nkhoma said, explaining that the Material Girl had kindly footed the bill for the entire village. "I hear she's divorcing her latest husband. Is that true?" I had no idea, but apparently it was.

Optimism abounded in Mwandama. New mud-brick houses bore shiny corrugated metal roofs. Bags of hybrid seed—donated by Monsanto, though not GMO—were piled high on the concrete floor of new brick granary the size of a high school gym, along with neatly stacked bags of fertilizer, also donated to the program. It was Thursday, which was now bank day in the village, and a bright-red armored truck from the Opportunity International Bank of Malawi was parked under a shade tree—part of the program aimed at providing low-interest agricultural credit to village farmers. A long line of men and women stood patiently in a queue outside the truck's teller window waiting to make their deposits, while more than a hundred villagers sitting on the ground nearby listened to a young banker explain how to get a loan.

Cosmos Chimwara, a 30-year-old cabbage seller wearing a Millennium Villages T-shirt, was standing in line to make his weekly deposit. I asked him how the cabbage business was going.

"The cabbage business is very poor, but it is going well," he said with a big smile. "I started with 26,000 kwacha (about US$185 in 2008), and now I have 35,000 kwacha (about US$250). I buy from the farmers and sell in Thondwe market. I go to farmers and buy cabbages at the garden price, then sell at the market price."

"Has having the bank truck helped?"

"My business started before the bank. The bank just assists us to have enough capital. Now I have three bikes, a TV, a mobile phone, and a better house."

Mary Austin, 49, was a few places behind him in line. She was a fish seller, traveling the 80 kilometers or so to Lake Malawi, where she would buy dried fish and bring them back to the village to sell. "It is a very good business," she said, holding her deposit of 500 kwacha in her hand. "This bank is very good to us. We know our money is safe."

Faison Tipoti, the village leader, sat in the shade of the corn-crib beside his house and smiled at the pleasant scene before him. Tipoti was a retired extension agent for the Ministry of Agriculture, and it was primarily his agricultural and organizational skills that had brought the famous project to his village. He was thick and muscular, with a kind, round face and twinkling eyes.

"When Jeff Sachs came and asked, 'What do you want?' we said not money, not flour, but give us fertilizer and hybrid seed, and he will do a good thing," Tipoti said in a thoughtful basso profundo. No longer do the villagers spend their days walking the road begging others for food to give their children with swollen bellies and sickness, he said. No longer must they drink from muddy open wells. Tipoti pointed to the new village well, an impressive concrete slab upon which rose a meter-high concrete obelisk with a single silver tap—not terribly different from something you'd see in Central Park. It was surrounded by children who happily splashed each other as they gathered water and washed clothes. "With the coming of the project, everywhere is clear, fresh water," Tipoti said. "Children are not born with dysentery and malaria. We are seeing our children grow up healthy and strong."

We drove over to the nearby "Justin" village, though Nkhoma did not know whether Justin's last name was Timberlake or something else. There we met 76-year-old Irene Maganga who wore a bright-yellow polo shirt with the number 87 emblazoned on her chest, along with a belt of five or six children clinging to her waist.

These were orphans she had adopted; they had lost their parents to the HIV/AIDS epidemic that plagued a generation of Malawians, with infection rates as high as 26 percent of the adults in 1998. Thanks to enormous efforts of the government and aid agencies, the infection rate has fallen to about 10 percent today.

Maganga was brimming with smiles, showing the few teeth she had left, as she described her new house, her Chichewa words punctuated with happy, ululating "Eeeeeeeeeee!" "These are the years I am enjoying the best," she told me through Nkhoma. "I never thought in my life I would live to be food secure." Her use of the NGO buzzword seemed a bit bizarre until she described just how important "food security" was to her. "The drought of 2005 was very difficult," she said. "We had to do piecework for the tobacco estates (that pays about 50 cents a day) and were eating one meal a day. Boiled mangoes, boiled cassava chips." At times she'd been so hungry she couldn't sleep. Even the men had suffered from malnutrition. "What can you do then but watch the people suffer?" she said. "But by God's grace, everyone in the village survived."

The coming of the Millennium Villages Project in 2006 was a godsend as well, she said. She went from harvesting 2 bags of maize—very low production even by Malawi standards—to harvesting 32 bags during the record crop of the previous year, though this year she had made only 17. After saving her money for four years, in 2008 she had built a new brick house, which cost her about US$400. It gave her status in the community and made her very proud. She could now pay school fees for some of the orphans, though not yet all of them. She had even bought her first bicycle—at age 73—though she assured me she'd learned how to ride one back in 1958.

OUR LAST STOP was at Mkumugwa village on a steep hillside that sloped down into a giant ravine. We toured a tree nursery where villagers were growing *Tephrosia vogelli*, a nitrogen-producing shrub that can be planted alongside crops to provide nutrients. But the most impressive project yet was just down a narrow dirt path, where a local village irrigation committee had hand-dug a rectangular pond, roughly 10 meters wide and 50 meters long, into the hillside. Fat tilapia weighing a kilogram or more roiled up to the surface of the muddy brown water to gobble up a handful of fish pellets. The men said they had eaten a few of the fish, just to see how they tasted, but were waiting for them to grow a little more before they sold them in the market.

Just downhill appeared an African mirage. A lush green field of sweet corn spread like a 10-hectare emerald floating in a sea of red, black, and brown, made possible by gravity-fed irrigation water from the tilapia pond. It was the only green field for as far as I could see, since no crops are grown in the dry season without water. The tall green stalks showed what Africa could do if the green revolution ever took hold. Borlaug's hand was evident in the farmers' use of his "Sasakawa method" of seed spacing for planting maize. He would have been pleased at the results.

In 2010, Pedro Sanchez and his colleagues published their data on the yield increases observed in eight of the Millennium villages. The study, which covered 52,000 farming households that included 310,000 people, showed that 75 percent of the farms had maize yields of 3 tons per hectare, triple the regional average. Fewer than 10 percent of the households had yields lower than 2 tons per hectare. The data clearly showed that when African farmers had the same basic inputs that are available to most farmers around the globe, they could grow a lot more food. And not just in Malawi. Africa has more than 300 million hectares of fallow arable land, along with plentiful surface water and groundwater that could be tapped for irrigation. Many experts believe that if Africa ever meets its agricultural potential, it, like Ukraine, could feed not only itself, but much of the world as well.

Malawi still has a wide gap to close. And just as the green revo-lution never made it to Africa, neither did Malthus's theory on fer-tility rates falling with real income. Large families have long been a source of pride and joy in Africa, as well as farm labor and social security for aging parents. Malawian women continue to have, on average, six children, and the United Nations has projected that the 2014 population of 16 million will more than triple to 50 mil-lion by midcentury. I asked Nkhoma if she thought Malawi would be able to feed itself as its population continued to grow.

"Look what we've done in just three years!" she said, unfazed. "We've gone from depending on aid to selling maize to other coun-tries. We have the Shire River flowing through half the southern region. We could irrigate all that area along the river. I think we will be able to keep up with population growth." Almost on cue, the clouds opened up and the first tentative rain of the rainy sea-son began. It wasn't a soaking, planting rain. But it turned the tour back to the truck.

As we climbed the steep hill toward the road, we stopped to catch our breath along the path. The rain shower had passed, and a dozen or so schoolchildren in their crisp white blouses and blue skirts sat in a nearby field on top of one of the rows that descended the field like bleachers. Nkhoma nodded to a little girl with two plaited pigtails who looked to be about seven, the age of my oldest daughter at the time. She asked her name, and the little girl told her, but she spoke too softly for me to hear. Then Nkhoma asked her what she wanted to be when she grew up. The little girl thought about it for a few seconds and then pointed to the heavens.

We both whirled to look, and we saw the faint silver speck of an airliner dragging a contrail across the sky. "A pilot?" The girl nodded. Nkhoma smiled. "Then you must study very hard."

Besides providing bed nets and water pumps, the Millennium Villages Project had fostered higher aspirations for at least one young girl in the village. And I got the feeling that such dreams

were not beyond her reach. But I couldn't help wondering what
would happen to her when the project ended. Or to the 7.5 million
other children in Malawi for whom that dream remains sheer fantasy. Who will foot the bill for them?

THE ROAD TO Ekwendeni lies in the exact opposite direction from
the road to Zomba, both physically and philosophically. It climbs
north along the steep spine of the Viphya Mountains, where it
courses through mile after mile of dense forest—mostly tree
plantations owned by foreign firms. It eventually spills down the
mountain to the northern Malawi town where Presbyterian mis-
sionaries from the Free Church of Scotland built a small hospital
more than a century ago. The hospital has a critical nutrition ward,
and during the hungry months from December to February, particularly in drought years, the clean, orderly cots are full of tiny
children with swollen bellies and reddish hair—the classic signs of
kwashiorkor, intense malnutrition.

Rachel Bezner Kerr, a young Canadian soil scientist working on
her master's degree at the University of Guelph, came to Ekwendeni in the late 1990s planning to study the potential impact
of small family vegetable gardens on increasing nutrition. After
meeting with health workers at the hospital and interviewing more
than 50 families, however, Bezner Kerr became convinced that the
community's nutritional woes stemmed directly from the extensive maize monoculture that provides, by some estimates, up to 75
percent of Malawi's calories.

With the help of the hospital and small grants from the Presbyterian Church in Canada and other nonprofits, she started a program to diversify the local farms with legumes, such as soybeans,
groundnuts, and pigeon peas, that were good for both the soil
and children's diets. But instead of the top-down approach used
by Borlaug, in which trained agricultural extension agents teach

local farmers how to use modern seeds, fertilizers, and pesticides, Bezner Kerr worked from the bottom up. She invited 30 farmers, both men and women, married and unmarried, from seven extremely malnourished villages around Ekwendeni to form a Farmer Research Team. Each farmer chose two different rotations of legumes (pigeon peas, groundnuts, or soybeans) and maize to test in a small section of their plots to see which one worked best, and they were then given a small amount of legume seeds and training on how to grow them.

The first year, 183 farmers joined the project, which was dubbed "Soils, Food, and Healthy Communities" (SFHC). But that number began to rise as their neighbors saw the results. Farmers trained their peers during field days, while experienced legume growers took on apprentices. Since most legumes self-pollinate (as opposed to the hybrid maize, which is cross-pollinated), poor farmers can save their seeds for replanting.

"It was not an easy transition," Bezner Kerr says of the project's humble beginnings. "It took a lot of farmers about four years to see real improvement. And there was a lot of skepticism. They'd been told for 30 years that you needed fertilizer to get better yields." There were also gender bias and village politics to overcome. To get the greatest benefit from legume crops, the stalks need to be incorporated into the soil after harvest, which meant an additional task for the growers. Since the project leaders did not want to add to the already heavy workload of the women in the program, they sponsored incorporation days to get men involved.

By 2012, more than 6,000 farmers were participating in the effort, and since most farmers support a family of five, some 30,000 people were affected, rivaling the Mwandama project in size. The similarities stop there, however. SFHC works on a fraction of the Millennium Villages budget (from 2000 to 2012, SFHC spent about $8 ($10 Canadian) per person per year, compared to

$160 per person in the Millennium Villages Project). I found Boyd Zimba, the project's assistant coordinator, and Zaharia Nkhonya, food security coordinator, sharing a one-room, window-cooled office in a wing of the hospital, whose array of yellow one-story buildings looked less like a medical facility than a well-tended kids' summer camp.

Zimba, Nkhonya, and I hopped into the project's battered pickup, only to discover the battery was dead. But with a good push and a popped clutch, we soon got it going and were rattling down a dirt road toward an SFHC village as the two men talked about the downside of Malawi's famed fertilizer miracle, which in 2008 had grown to $250 million, nearly 15 percent of the government budget.

"First, the fertilizer subsidy cannot last long," said Nkhonya, a compact man with a ready smile who was well aware of the ballooning cost of the government program. "Second, it doesn't go to everyone. And third, it only comes once a year, while legumes are long-term—soils get improved with organic matter every year. It's highly sustainable. The fertilizer subsidy is political. If this president goes out, the next one might not continue it."

We stopped in the tiny village of Encongolweni and were greeted by a chorus of 17 SFHC farmers, both men and women, singing one of the project's songs about the dishes made from soybeans and pigeon peas. There are no rock stars here. No Madonna Village. After the introductions, we all sat down on the dirt floor of the village's thatched-roof meetinghouse as they got up, one by one, and testified about how legumes had changed their lives.

Ackim Mhone, a spare farmer who looked to be in his 30s, had joined the project in 2001, and his experience was common among the program's participants. "Before I joined the project, I would harvest 10 bags of maize," he said. "After I joined, I got 20–25 bags of maize from the same piece of land. That was enough to change the life of my family. Before the project, my family was getting sick now and then, mostly the children. But we are able to go year-round without getting sick now. I now have a push bike and a brick house. I now have goats and pigs."

Esnaly Ngwira's story was the most impressive. Before joining the program in 2001, the 54-year-old widow had had very little food for her family. She started growing groundnuts in 2001 and incorporating the residue into the soil. The next year she noticed her maize crop was growing vigorously. In 2004 she planted half an acre of groundnuts intercropped with pigeon peas. She incorporated the residues again and planted maize the following season. In 2005, when most maize fields in Malawi were scorched by drought, she harvested her best crop ever. "That same year my husband passed away," she said. "But we could still feed ourselves and live okay." She can now support her six children and two grandchildren, whose parents have passed away.

After all the farmers had told their stories, we toured their communal seed house. It was tiny compared to the giant warehouse at Mwandama—a single room roughly 10 meters by 20 meters that the group itself had built. But it was stout, with brick walls, a concrete floor, and a corrugated steel roof. There were no bags of donated Monsanto maize here, or pallets of donated fertilizer. Just a hundred or so neatly stacked bags of legume seeds awaiting the next harvest. And that was one of the challenges. Far more farmers want to join the SFHC project than there are legume seeds to provide them.

This year the farmers were diversifying even further. The group had purchased tomato and onion seeds, which the farmers planted in a low field near a year-round spring they used for irrigation. It was their first attempt at growing crops during the dry season to sell at the market, and the vegetables were thriving. Alice Sumphi, a 67-year-old farmer whom Bezner Kerr remembers being extremely food insecure when she joined the program in 2000, was now literally dancing in her tomato field, laughing and pointing out that her knee-high plants were far bigger than those in the plots of younger men nearby.

As we walked up from the tomato fields, Sumphi took me by a makeshift corral and proudly showed off a large pink sow and its piglet. She'd bought the mother pig from the profits she'd made from farming and had sold more than 20 piglets for cash. "What's her name?" I asked Sumphi. "Kudya!" she said with a laugh—spitting out the Chichewa word that means "to eat"—"because she is always eating!"

Back at the hospital, I visited the nutrition ward to see whether any children were present. Even though the hungry season had begun, the neat rows of small beds were all empty. After seven years, as farmers in the program boosted yields and diversified their diets with legumes, their children were experiencing significant gains in both height and weight, particularly in families that had been growing legumes the longest—making a convincing case that soil health and community health are related in rural Malawi.

Malawi's fertilizer subsidy program, on the other hand, nearly bankrupted the country. In 2010 the IMF and the World Bank asked President Mutharika to devalue the kwacha in order to meet the requirements for further funding. Doing so would have exploded the price of imported fertilizer and ended the subsidy, the key to his political power. He refused, and this time the bankers pulled their funding. The economy went into free fall. In July 2011, tens of thousands of Malawians took to the streets to protest government corruption and high prices. The army shot 19 of them, killing 11. In 2012, with his country's economy in shambles and protests mounting, Bingu wa Mutharika, the father of Africa's modern fertilizer subsidy programs, died of a heart attack. That year the maize crop was down 40 percent because of erratic rainfall, and the next year maize prices more than doubled—requiring large, back-to-back food lifts from the World Food Programme to stave off disaster. Nearly 4 million people throughout Malawi needed food aid in 2013—the largest relief effort in a decade.

Jeffrey Sachs was about the only one who mourned Mutharika's passing, praising him in the *New York Times* for standing "bravely against the arrogance of an ill-informed foreign-aid community" by reinstating the fertilizer subsidy. "Lo and behold, production doubled within one harvest season. Malawi began to produce enough grain for itself year after year."

Only, it turns out, they actually didn't. The same year (2007) that Malawi reported its record 3.4-million-ton maize harvest, one of the country's steadfast donors, the government of Norway, sponsored an agricultural census by Malawi's National Statistical Office that interviewed thousands of farmers across the country to collect yield data. The government of Malawi buried the report until 2010, and dismissed it when it came out. No wonder. The Norwegians and their colleagues in the statistics bureau calculated the 2007 harvest at only 2.1 million tons—the minimum required to keep Malawi from starving. If the Norwegian numbers were accurate, the Ministry of Agriculture had inflated its maize yield figures by 40 percent. Moreover, the government's estimates were nearly double the Norwegian figures for rice, fourfold for millet, fivefold for sorghum, eight times those for cassava, and fully nine times higher for sweet potato. While Mutharika was being lauded around the world for feeding his people, several aid experts noted that the price of maize hadn't fallen at all because of the pro-claimed surplus, and many people in the country were struggling to keep from starving. The Malawi Miracle wasn't so miraculous after all.

"It was a great media blitz," says Thomas Jayne, an agricultural economist at the University of Michigan who has studied Malawi's

subsidy program, as well as others in Africa. "There's no denying the subsidy increased maize production, and most of that was
consumed by the people who grew it. But it did not increase food
security in a real sense. For $250 million a year, you'd better be
getting some benefit." More disturbingly, he says, the program has
been plagued by corruption, with as much as a third of the fertilizer being sold on the black market to larger, wealthier farms and
political elites. It's done nothing to reduce rural poverty. When
the program began in 2005, 50 percent of rural Malawians lived
below the poverty rate. By 2010, their number had increased to
53 percent.

The Millennium Villages Project (MVP) has had trouble with
numbers as well. Sachs has long refused to release hard data on
the program and has published few evaluations of the project in
peer-reviewed journals. In May of 2012, however, MVP researchers, including Sachs, published a paper in the medical journal the
Lancet reporting that the child mortality rate had fallen three times
faster in the MVP villages than elsewhere—implying that their
big-push intervention was having a tremendous impact. Unfortunately, their math was as bad as that of Malawi's Ministry of Agriculture. The corrected numbers showed that the under-five child
mortality rate in the MVP villages over four years had actually
fallen less than the average national child mortality rates reported
by the nine countries that have Millennium villages. Though statistically insignificant, the difference implied that children had a
better chance of survival outside the villages than in.

No one doubts that the Millennium Villages Project and fertilizer subsidy programs have helped hundreds of thousands of Africans. The real question is, Do they provide the best bang for the
millions of bucks they cost, or could that money be spent better
elsewhere? Even less expensive programs like the Soils, Food, and
Healthy Communities project struggle to maintain their gains. In

2009, researchers conducted a random survey of 128 participants in the program, most of whom had been involved for six years or more. Only about half were still growing legumes, despite the documented benefits they provided. Many had been seduced away by subsidized fertilizer and had returned to planting only maize. Both the MVP and the SFHC experiments suggest that there are no easy solutions to Malawi's or Africa's food problems.

Successful farms were not hard to find, however. Leaving Ekwendeni, I asked my taxi driver to take the road along Lake Malawi back to Lilongwe. When we reached the shimmering water, I was surprised to see mile after mile of dense, green, irrigated sugarcane fields lining the edge of the lake. The driver told me the plantation was owned by Germans and had been there since colonial days. Such large plantations occupy 16 percent of Malawi's arable (and best) land and produce sugar, tea, and tobacco for the international market. If President Mutharika had been serious about growing more food for his people, he could have grown a lot of it on plantations currently dedicated to exporting sugar. Land reform efforts have accomplished little.

Such large corporate farms are the latest—and most controversial—agricultural trend to hit Africa's shores in decades. Since 2007, the record prices of corn, soybeans, rice, and wheat have ignited a global land rush by private investors, particularly in countries where leasing or buying acreage is cheap, governments are amenable to deals, and property rights are frequently ignored. Malawi's eastern neighbor, Mozambique, is a hotbed of such large land deals, leasing millions of hectares to foreign corporations, from as far away as China and Brazil, that promise to build roads, irrigation, and infrastructure, as well as to bring in inputs like pesticides, fertilizer, and equipment and train small farmers how to use them. After 30 years of failure, corporate agriculture—enticed by high food prices, booming economies (6 of the world's 10 fastest-growing economies are in Africa), and peaceful stability—may finally bring the green revolution to Africa.

"If you wrote a letter to God and asked him for the best soil and climate conditions for farming, this is what he'd send you," says Miguel Bosch, an Argentine agronomist who is trying to start up a 10,000-hectare soybean farm near Gurúè, Mozambique, called "Hoyo-Hoyo," on land that was once a battleground in that nation's 20-year civil war. Now, Bosch says, it's a paradise for growers. "I've spent many years farming in Brazil and Argentina and have never seen such soil."

Unfortunately, the farm's previous managers failed to deliver the promised schools, clinic, or land to the smallholder farms they displaced, creating a backlash from local farmers. Much of the farm's new equipment has been sabotaged. A Chinese corporate farm ran into similar problems in the Limpopo delta when it displaced thousands of small farmers to renovate an old Soviet-built irrigation system to grow 20,000 hectares of rice. Even so, the government is banking on such projects to dramatically increase food production in the nation. Mozambique is currently working with the governments of Brazil and Japan to develop a North Carolina–sized swath of the country into a soy-growing powerhouse—similar to the development scheme that turned Brazil's Cerrado into the soybean basket of China. Officials envision a corridor dotted with 10,000-hectare corporate farms run by Brazilian companies, with training centers to help local farmers boost their own production.

So far, the ProSavana project remains little more than a pipe dream. But when small farmers and corporations have worked together, good things have happened. Elsewhere in Mozambique, small and middle-sized farmers are growing soybeans on their small plots under contract to larger companies, which are selling them to the country's booming poultry industry—which also contracts with small chicken farmers. Other contract farmers are growing organic fruits and vegetables for export to Europe.

A few critical differences make Mozambique's prospects far brighter than Malawi's. It's a coastal nation, with the best natural harbors on Africa's Indian Ocean coast, making trade with the rest of the world easier. But much of the nation's economic and agricultural boom is being driven by a huge natural-resource boom. Giant new mines are digging into one of the world's largest untapped coal deposits, ramping up demand for labor, wages, and food. New, world-class natural gas deposits have been discovered along its northern coast, along with other minerals and precious metals, leading to tremendous foreign investment and donations by countries wanting to curry favor with the government. China has built two major airports, the nation's parliament building, a national soccer stadium, and most recently, an opulent presidential palace. Most important, Mozambique has millions of acres of fallow land and one of the lowest population densities in Africa. Average smallholders there farm 1.5 hectares, almost double the average farm size in Malawi.

To some, this sounds like an environmental catastrophe. Humanitarian groups call corporate land deals "landgrabs" and "agroimperialism." Yet veterans of agricultural development say the massive infusion of private cash, infrastructure, and technology that such deals may bring to poor rural areas could be a catalyst for desperately needed change—if big projects and small farmers can work together. The key, says USAID's Gregory Meyers, is protecting the land rights of the people. "This could significantly reduce global poverty," Meyers says. "And that could be the story of the century."

THERE IS, OF course, another way to reduce poverty and hunger. One study that appeared in 2012 attracted hardly any attention at all but has implications just as profound for the continent's food prospects. A group of researchers from the Futures Group, a global health consultancy based in Washington, DC, analyzed three sophisticated projections of food security in Ethiopia in 2050. Like Malawi and Mozambique, Ethiopia is a country in the crosshairs of climate change. Average temperatures are rising, rainfall has become more erratic, and maize yields are projected to decline steeply in the coming decades. By 2050, Ethiopia's population will more than double to somewhere between 154 and 194 million people.

Already, 60 percent of the population eats 220 calories a day less than the amount that the World Health Organization recommends for good health and nutrition. If Ethiopia hits the high population mark, the researchers calculated that the country's caloric deficit will more than double, to 500 calories per person a day. Such a drop in caloric intake also would likely lead to greater poverty, poorer health, and more angry, unemployed young men facing a future of misery and strife.

On the other hand, if Ethiopia hits the low mark, which would require families to have an average of two children instead of five, population will peak at 2050 and then start a gradual decline. That scenario would cut the average per capita caloric deficit for Ethiopia almost in half, to just 127 calories a day by midcentury. "This reduced shortfall is equivalent to the shortfall in the medium fertility scenario without climate change," the authors reported, "meaning that a low fertility scenario has the potential to fully compensate for the impacts of climate change on food consumption."

And Ethiopia can get there, simply by expanding contraceptives and family-planning services to meet current demand. According to surveys, a quarter of Ethiopian woman would like to use contraceptives but can't access or afford them.

Reducing population growth has long been the third rail of development work, steeped in cultural, religious, and political controversy. Few are willing to even talk about it, much less advocate it. Yet with climate change threatening to slash yields on a continent already suffering a tremendous food deficit, the subject can no longer be ignored. Africans need access to birth control as dearly as they need higher crop yields.

"When people talk about adapting to food scarcity caused by climate change, they almost always focus on improving agricultural output; they rarely come back to population as an issue that can be addressed," said Scott Moreland, the lead author of the study when it was released. "Improving the food supply is an important concern, but people can sometimes overlook the demand side as well."

Which brings us, inevitably, back to Malthus.

An image from Yann Arthus-Bertrand's exhibit *Earth from the Air*, on display outside Bath Abbey in 2010.

CHAPTER 14

The Grand *Desiderata*

We are not, however, to relax our efforts in increasing the quantity of provisions, but to combine another effort with it; that of keeping the population . . . at such a distance behind as to effect the relative proportion which we desire; and thus unite to the two grand *desiderata*, a great actual population, and a state of society in which squalid poverty and dependence are comparatively but little known.—*T. R. Malthus, 1803*

The train from London's Paddington Station to Cardiff rolled out of the great city on a bright February day and was soon clacking past hedge-bordered fields of gray English countryside. I lost track of time staring at the farms and villages. It was in just such a setting two centuries ago that a mathematically gifted young minister counted far more baptisms than funerals in his tiny rural chapel, saw the peasant children stunted from hunger, and deduced that humanity's ability to produce children was far greater than our ability to feed them. Most galling to the utopian writers of the day was Reverend Robert Malthus's assumption that humans, despite our science, culture, and reason, were ultimately bound by natural law. Although his stock has risen and fallen over the centuries, Malthus's basic challenge to the world remains. We are locked in a never-ending two-step between our numbers and the sustenance we can wrest from 6 inches of topsoil.

It seems fitting that the man who tried to understand humanity's ebb and flow would be buried in Bath, a city that has overlooked the lush green valley of the river Avon since Roman legions conquered the land. I hopped off the train at Bath's station and, after a brisk walk along cobblestoned streets, was soon standing in front of the town's castle-like abbey, a fluted monolith of honey-hued limestone. Fat cherubs climbed stone ladders up the façade to the bell tower, while a gnarled olive tree—a sign of Earth's abundance—sprouted from one corner in bas relief. The structure, begun in 1499, was one of the last great medieval churches built in England. It rose from the ruins of a previous cathedral built by Norman conquerors 400 years earlier, a church so large that the current abbey, which seats 1,200 people, would have fit nicely in its nave.

In Malthus's day, the city thrived as an epicenter of society and art. Jane Austen lived here for several years, skewering Bath high society in two novels, while Charles Dickens wrote for the local paper. Young Robert Malthus went to boarding school at nearby Claverton Manor and was living in Bath when his son was born. He had several friends and relatives here that he visited throughout his life, including his wife's parents, whom he adored. After one visit to his in-laws' house around Christmas of 1834, the 68-year-old Malthus suddenly took ill. He died the next day and was buried in the floor of the abbey along with countless other luminaries who had come to be healed by England's sole natural hot spring—only to die here instead. His large memorial tablet, cemented in a prominent spot on the wall of one of the main entrances to the church, is passed by hundreds of tourists and local parishioners every day.

I wanted to see whether anyone in Bath remembered the town's infamous adopted son, whom his enemies savaged as "that black and terrible demon that is always ready to stifle the hopes of humanity" (Godwin)—the man who wrote "this vile and infamous doctrine, this repulsive blasphemy against man and nature" (Engles), this "gospel of despair" (Boner). His friends, of course, remembered Malthus quite differently. In faded letters carved in hard-to-read script, his memorial stone celebrates "the spotless integrity of his principles, the equity and candour of his nature, His sweetness of temper, urbanity of manners, and Tenderness of heart, his benevolence and piety are the still dearer recollections of his family and friends." No one I spoke with in Bath remembered him at all, including a sixtyish tour guide outside the church. "Mowfus?" he replied to my query. "Never 'erd of 'im."

I spent hours wandering around the abbey, dubbed the "lantern of the West" for all the afternoon light that poured through its many windows, looking for Malthus's actual grave. So many dignitaries have been buried here over the centuries that you cannot walk down the aisle in any direction without defiling a duke, a general, or even one US senator. I never found him. I later learned that his grave had been covered by a pew. Such is the ignominy of time.

Outside the church, however, he was everywhere. The plaza surrounding the abbey was a popular place for public art, and by chance, the day I visited, the building was encircled with an outdoor exhibit of *Earth from Above*, the mesmerizing aerial images by French photographer Yann Arthus-Bertrand who has documented both Earth's natural beauty and our heavy human footprint upon it. The photographs, enlarged to the size of small billboards and

intricately detailed, drew the viewer in. Here was population over-
flowing in the sprawling floating slum of Makoko in Lagos, Nigeria,
where some 80,000 people—a fraction of Lagos's 21 million—
inhabit a leaky fleet of stilted shacks moored in a polluted lagoon
with no electricity or running water. Here was agricultural production, in the form of a giant feedlot outside Bakersfield, California,
where multicolored cows stretched to the horizon. Here was the
deep black scar of an open-pit coal mine in South Africa, the well-
spring of climate change and the industrial revolution. This last
image stood beside another depicting endless plowed fields on Turkey's Anatolia plain, cultivating green-revolution monocultures of
wheat and barley. Our ancestors domesticated wheat in southern
Turkey more than 10,000 years ago. Yet if current temperature
trends continue, it likely won't survive there by century's end.

The exhibit concluded with that old time line that condenses
all of Earth's history into a single year. Modern humans crash the
party an hour before midnight on New Year's Eve and, seconds
before the ball drops, drastically alter the planetary systems that
sustain us. In just the last 200 years of the hydrocarbon-based
industrial revolution, we've changed the basic chemistry of the
atmosphere as well as the oceans; our bottomless demand for food,
water, energy, and space shoves thousands of our fellow species off
Earth every year, launching what Harvard biologist E. O. Wilson
calls the planet's "sixth great spasm of extinction." It's no coincidence that in the four decades since Norman Borlaug won the
Nobel Peace Prize, the population of other vertebrate species—
mammals, birds, reptiles, fish, and amphibians—has plummeted
by more than half. If we continue at this pace, one day the next
species we extinguish may be our own.

Of course, it doesn't have to end that way. Unlike other species,
as Malthus recognized in the 1803 revision of his famous essay,
humans have foresight. We can change and, more important, we

have changed our behaviors when food or wages were tight by practicing what he called "moral restraint," marrying later and having fewer children. Many modern critics accuse Malthus of predicting that populations would ultimately outrun their resources and then collapse amid war, famine, and disease. In reality, the father of demography spent his life trying to understand why populations *didn't* collapse.

Malthus disliked extremes. He strove to find the balance point in any given system that would give the most benefits to the most people. He advocated for tariffs on cheap imported corn—going against David Ricardo and other free-market economists—because he did not think England should depend on foreign nations for its food supplies (sound familiar?). He was a lifelong proponent of state-run universal education for the poor and of giving the vote to landless laborers, to get them vested in the economy and moving up the economic ladder, and for delaying marriage until men could feed, clothe, and educate the children they had.

Today we find ourselves caught in Malthus's vise. Since the food price crisis began in 2007, agricultural experts have focused primarily on the need to double food production. We continue to debate the best way to do this. Jon Foley, former head of the Institute on the Environment at Borlaug's alma mater, the University of Minnesota, advocates what he calls the "silver buckshot" approach. His five-point plan includes these steps: (1) Stop farmland expansion, especially in the tropical rain forests. (2) Use existing agronomy to close yield gaps in Africa, Latin America, and eastern Europe. (3) Reallocate critical inputs like irrigation water, fertilizers, and chemicals, from places where they are overused to places where they are scarce. (4) Shift our diets away from meat and wean our cars off biofuels. (5) Reduce the amount of food that is discarded, spoiled, or eaten by pests, which amounts to fully a third of global agricultural production. Entirely eliminating waste from farm to fork could make 50 percent more food available to the planet.

It sounds simple, doesn't it? But each item on Foley's to-do list is daunting. The world has struggled to solve many of them for decades. Before the green revolution and the widespread use of pesticides, farmers lost about 30 percent of their crops to weeds, insects, and disease. Today, despite our far higher levels of food production and multi-billion-dollar "crop protection" industry, we still lose about 30 percent of our crops to such pests. We've struggled to close yield gaps in the developing world since the 1950s, to halt deforestation since the early 1990s (with some success), to eat less meat since medical studies confirmed J. I. Rodale's health fears in the 1960s.

"Providing food and nutrition for 9 billion people without compromising the global environment will be one of the greatest challenges our civilization has ever faced," Foley said when the study was released. "It will require the imagination, determination and hard work of countless people from all over the world, embarked on one of the most important causes in history. So let's work together to make it happen. There is no time to lose."

Reducing demand, while never easy, is the low-hanging fruit—far more within our control than doubling global food production. Yet it's telling that only one of Foley's bullets is aimed at demand. If the governments of the United States and Europe reversed course on food-based biofuels, 38 percent of the US corn crop and 16 percent of the European Union's arable land could be returned to growing food. The demand for meat is expected to double by 2050, but it doesn't have to rise that much. People in the United States have the highest meat consumption on the planet, averaging 322 grams each day. That's roughly the equivalent of eating a quarter-pound hamburger at breakfast, lunch, and dinner. So unhealthy is the average high-meat, high-fat, high-sugar Western diet that researchers believe it could shave five years off expected life span in the United States. If consumption trends continue, by 2020 half of all American adults could have diabetes or pre-diabetes, costing the nation a *half trillion* dollars each year.

A recent study in the United Kingdom estimated that if people there cut their meat and dairy consumption in half and replaced those calories with fruits, vegetables, and cereals, the effort would decrease greenhouse gas emissions by 19 percent and save up to 43,600 lives from diet-related diseases each year. Another study along the same lines from the University of Exeter suggested that if consumers in developed countries halved their meat consumption, recycled more animal and crop waste, and used bioenergy crops more efficiently, we would have enough land to produce food for 9.3 billion by 2050 without destroying more forests. More important, since both cattle and the fertilizer needed to grow feed corn are huge sources of greenhouse gases, a low-meat diet combined with more efficient agriculture could cut 25 parts per million of carbon dioxide from the atmosphere, possibly enough to keep us from crossing the 2-degree climate tipping point.

Of course, the most beneficial thing humanity could do is to show more restraint in the bedroom—a subject just as controversial now as it was in Malthus's day. The minister wrote candidly about "the passion between the sexes" being necessary and "a fixed law of nature." The preacher in Malthus condemned sex before marriage and, unlike today, laid the onus strictly on *men*. *Men*, he argued, must practice moral restraint and not engage in sex before marriage, because that could lead to children that they could not support. *Men* should wait to marry until they can support children, or suffer harsh economic consequences.

The United States and England have led the international family-planning movement since the early twentieth century, pioneered by Margaret Sanger in the United States and Marie Stokes in the

United Kingdom, among others. The movement blossomed in the 1950s and 1960s because booming populations were exacerbating food stress, particularly in India. Some places took the movement too far. India forced sterilizations on people (as did many US states, including my home state of North Carolina, on those it deemed unfit to breed), while in 1979 China instituted its infamous "one-child policy," after drought and Mao's poor agricultural policies starved much of the country during the 1960s.

China never actually got one child from the one-child policy. Rural areas and minority groups were not required to comply. But since women can give birth for roughly 30 years, there is a classic demographic delay: death rates fall first, followed by fertility rates, and eventually population. Even though China's fertility rate actually started falling in the 1950s and the rate is now thought to be 1.4 or lower, so large was the starting point that the actual population won't start dropping until sometime in the 2030s. The fertility rate is now so low that it will lead to a rapid aging of China, followed by a rapid decline in population. China's population is actually projected to plummet by 400 million people by 2100.

During the 1970s and 1980s, several economists in the United States, led by the University of Maryland's Julian Simon, turned the population debate on its head, arguing that from an economic standpoint, people were the "ultimate resource." The more people there were on Earth, the more productive they would be. Such arguments, along with the increasing controversy over government-funded abortions overseas, led the United States to sharply reduce funding for international family-planning programs during the last two decades of the twentieth century. Other nations reduced their contributions to the movement as well. From 1995 to 2007, international support for family-planning efforts fell by more than half. International aid was shifted to fight AIDS, deforestation, and other environmental concerns.

In Malthus's day, kings and regents and tsars wanted a large population to fill the ranks of their armies and provide abundant and cheap labor for the fields. A large population was deemed a sign of a country's strength, no matter how much misery it caused the poor. But we now understand the tremendous benefits that a slowly falling population brings. During the early 2000s, two Harvard researchers, David Bloom and David Canning, as well as others, began looking at the rapid rise in per capita income in China, India, and several other "Asian Tiger" countries that occurred during the 1980s and 1990s. Much of that tremendous economic growth, they concluded, came from a large bulge of young workers moving into the workforce who were supporting a smaller group of elderly dependents and young children, thanks to falling fertility rates. Governments could take money that they would have spent on education or on health care for the elderly and invest it in infrastructure, research, and manufacturing that drove economic growth. Bloom and Canning called this windfall the "demographic dividend." But in order to claim it, countries first had to go through the "demographic transition." That's the decline in fertility rates from more than 5 to about 2, with 2.1 being the replacement rate and 1.9 or 1.8 being ideal to provide a slow, smooth curve that gradually decreases the numbers of the very young and the very old while providing a large productive workforce in the middle that can support them. The demographic transition is also a bellwether for something even more important than money: democracy.

During my visit to England I swung by demographer Tim Dyson's office at the London School of Economics, where he was working on a new paper on the subject. Dyson, who looks a bit like a disheveled Colin Firth in blue jeans and hiking boots, was sitting amid a sea of boxes and stacks of papers. He shuffled through a stack and produced a page showing graphs of the demographic transition in several countries around the world, including Spain,

Chile, Taiwan, India, and England. With a pencil he pointed to the graph of England's population over the last 200 years.

"That point, 1817, is the start of real democracy in England and Wales," Dyson said. "For three reasons. First, the rate of natural increase starts to fall because the birth rate has fallen, so the population starts growing slower. Second, women start to get more interested in things other than childbearing and start clamoring for the vote. Last, and most important in many respects, the population gets older. This also causes urbanization." Both urbanization and an older population are key for democracy to flourish, says Dyson. The demographic transition "is the most important thing that has happened in your country, in my country, in China and India and any damned country in the world during the last 250 years."

Sub-Saharan Africa has yet to get the memo. Of the 25 countries left on the planet where the average woman has five or more children, 23 are in Africa, with the other two being Afghanistan and East Timor. In none of these countries do women have equal rights; in many they are virtually enslaved. Yet their freedom—economic, political, and sexual—is the key to solving the global food crisis. African women spend 90 percent of their income on their families; African men spend 30 percent. Women farmers now grow nearly half of the world's food, including most of the food in Africa, yet because of their second-class status they have far less access to agricultural technology, tools, and bank credit than men have. A recent USAID report suggested that if African women were simply given equal access to agricultural extension services and farm credit, Africa's production would increase 30 percent overnight. If you added equal access and title to land, they would create nothing short of a "pink revolution."

Studies have also shown that girls who get at least seven years of schooling marry later and have far fewer children than those who don't. Universal education, as Malthus said, is among the best investments that governments can make.

Where governments have educated women and made family-planning services available to all, fertility rates have dropped quickly on their own, without draconian one-child policies or other harsh laws. A prime example is Iran. You read right. The Islamic Republic of Iran holds the record for making the fastest demographic transition, reducing its fertility rate from more than 5 to 2.1 or lower in less than 15 years. Iran's 6.5 fertility rate after the 1979 Islamic Revolution was even higher than in many African countries today. The high rate wasn't deemed a problem until after the bloody war with Iraq ended in 1988, when the country faced an extreme economic crisis. Iran's national budgetary agency informed the prime minister that the nation did not have the funds to rebuild its infrastructure and provide social services for its people.

Ayatollah Khomeini had actually approved contraception years before, but not all of the powerful Islamic leaders agreed with him. So a handful of concerned doctors and health officials went to the Supreme Judicial Council and argued that the Koran prohibited harming oneself, and that by having six children Iranian women were putting themselves at high risk of death during frequent childbirth. The judicial council soon approved the plan, and in 1989 Iran launched a broad new family-planning program aimed at reducing pregnancies in the early and late years of fertility, spacing out births, and providing a wide range of free contraception to all women throughout the public primary health care system.

The strength of that system in rural areas was key. In Iran, where half the population lived outside of cities after the revolution, each rural village has a "health house" staffed by a man and a woman from the village who go through an intensive but practical medical training for two years—the equivalent of an army medic or nurse.

The men are responsible primarily for maintaining sanitation and safe water supplies, while the women provide family-planning services. Both provide primary health care and vaccinations, with a focus on preventative care. Iran's health house system has been so successful that it has been replicated in many developing countries in Latin America, and has even been proposed for the Mississippi Delta in the United States, where the current health care system has utterly failed its impoverished population, despite millions spent on the problem.

Experts believe that at least a third of Iran's fertility decline came from making education available to girls. By the mid-1990s, women's literacy in the country had reached 74 percent and female employment had grown more acceptable. By the turn of the twenty-first century, there were more women in university classes than men. Iran's fertility rate reached the replacement rate of 2.1 in the early 2000s.

Unfortunately, Iran squandered its demographic dividend through poor political leadership that isolated the country from world trade and kept its economy in the tank. In fact, Iran's economic woes and related unemployment have become so bad that young Iranians are now delaying marriage even longer, dropping the nation's fertility rate lower than replacement levels. This trend led Iran's former president, Mahmoud Ahmadinejad, to encourage women to marry at age 16 or 17 and even to offer direct payments to newborns that would continue until they reached adulthood. Few women, however, took him up on the offer.

An increasing number of developing countries have joined Iran in completing the demographic transition to replacement fertility rates or lower, including Cuba (which achieved that rate by 1980), Brazil, Chile, Costa Rica, Lebanon, Myanmar, Thailand, Tunisia, and the United Arab Emirates, to name a few. Demographers are looking to South Asia and Africa next, though the push to reduce population growth, by necessity, must come from within. As climate change begins to hit the continent harder, causing increased food insecurity, African leaders may realize that the demographic transition provides a food dividend as well. This is becoming painfully apparent in the Sahel. The nation of Niger has the highest fertility rate in the world, at 7.6 births per woman, and ranks dead last on the 2013 United Nations' Human Development Index of the world's 186 countries. Even if Niger's fertility rate is halved by 2050, its population will nearly triple, from 17 million to 50 million people, in a nation that has already lost half its arable land to the encroaching Sahara desert. In 2013, 10 million people across the Sahel region were threatened with food insecurity, including a million children who were at risk of severe malnutrition. It should surprise no one that 10 of the 15 countries deemed at the highest risk for state failure by Sandia National Laboratories' Human Resilience Index in 2012, have some of the highest fertility rates in the world.

The greatest gift the developed world could give the developing world—as well as itself—would be to help poor, rapidly growing nations make family-planning services free and available to everyone. Even though large families are seen as a sign of wealth and prestige among African men, among African women it's often a much different story. Forty percent of all pregnancies in the developing world are unintended—75 million every year—roughly half of which end in abortions. Unsafe abortions cause an estimated 13 percent of maternal deaths each year, which should be considered a crime against humanity, since those deaths are entirely preventable through access to birth control, if not safe abortion clinics. Strong family-planning programs could help reduce the projected population of Africa by a *half billion* by 2050.

Don't get me wrong. Every child on Earth deserves to be fed, educated, and given the opportunity to reach his or her potential on this planet. Our challenge is to reduce our impacts, as well as those of future populations, so that "the goad of necessity" will not be so hard on them.

"Family planning is just about the only thing that works," says Tim Dyson. "It's cheap, it's effective, and it has huge beneficial consequences—slower population growth, slower growth of towns, less urbanization, and it liberates women. There's no better modern technology that I know of, and there are about 30 different versions of it as well." If all the world's 75 million unwanted pregnancies were prevented—meaning all women on the planet had access to family-planning services and free contraceptives and were able to have only the number of children they wanted—the average global fertility rate would drop below the replacement value of 2.1 almost overnight. This would lead to a global population of around 8.4 billion by 2050, falling to about our current level of 7.3 billion by the century's end . . . and 3.9 billion fewer people to feed than the current medium UN projections in 2100.

Helping poor, malnourished countries slow their population growth, however, is a long-term goal, and only half of the solution. It has to go hand in hand with a reduction in consumption of natural resources by rich, developed countries to avoid widespread ecosystem collapse. Few countries have made the transition to low birth rates without modern economic growth, which means increasing industrialization, urbanization, and fossil-fuel energy. The good news, in one sense, is that southern Africa, where the demographic transition is most needed, is at the dawn of an energy boom, with vast new oil fields discovered off Angola, new, world-class gas fields discovered in Tanzania and Mozambique, and one of the largest deposits of coal recently discovered in Mozambique as well. Amid all the calls for more expensive imported fertilizer to Africa, it is the ultimate irony that the continent contains half the world's remaining reserves of phosphate and vast deposits of natural gas, the primary feedstock of nitrogen fertilizer, giving that continent the raw materials for two of the three primary fertilizer nutrients.

In another sense, however, this is a disaster. To keep global emissions down to a manageable level, the developed world will need to offset Africa's and Asia's rising emissions, building a new path toward low-carbon-emitting economies that developing nations can follow. Yet despite more than two decades of warnings about the catastrophic consequences of climate change, our emissions have only increased. In fact, scientists believe we are now pumping carbon dioxide into the atmosphere faster than at any other time in Earth's long, 5-billion-year history. We are headed into uncharted waters or, as Timothy Dyson likes to say, we're on a roller coaster that has just left the tracks.

THERE ARE TWO paths before us. If trends continue, we will be buffeted by rising food prices resulting from each climate shock, be it drought in Australia, heat waves in Europe, or the dearth of irrigation water in California or the Great Plains. Grain prices will continue to rise, causing meat prices to soar. Fish and shellfish stocks will plummet, as a result of ocean acidification and warming, as will global biodiversity as forests and fertile plains fall to the plow. Meat at mealtime may return to a once-a-week treat instead of the daily staple it has become. Rising sea levels will increasingly flood rice-growing deltas, while governments will be faced with demands for expensive seawalls and dikes from hundreds of coastal municipalities, large and small. The economic triage will protect the most important urban areas—Washington, DC, New York, London, Los Angeles, Tokyo—leaving rural farmers to fend for themselves and head for higher ground.

One of the more wrenching challenges will come from immigration, as growing ranks of young, hungry, impoverished, poorly educated people in Africa and South Asia clamor to escape countries that can no longer clothe or feed them. Such immigration pressure is already apparent in the United States and Europe, because of the disparities in employment and wages. One can only imagine what it will be like when the driver becomes food.

This isn't some dystopian fantasy from a teenage novel. It's already under way. In March 2014 the International Panel on Climate Change issued its 32-volume Fifth Assessment—its starkest warning yet. The 300 or so authors concluded that climate change is already depressing crop yields, especially those of wheat and corn; fueling droughts, floods, and heat waves that hamper food production; and contributing to food insecurity, high prices, and conflict in many places around the world.

The other path, however, brings us a host of benefits. It would start by making education through high school a global civil right, freely available to all children. Perhaps we could use our incredible technology to create culturally appropriate courses, designed by and for developing countries in their own languages, and beam them to rural schoolhouses all over the planet. Next, we should put aside our political and religious differences and provide universal family-planning services, education, and contraception. We should strive to ensure equality for women in education, services, access to loans, land, and equal paychecks, to truly unlock the potential of half the world's people. We must invest in basic agricultural research and extension services to help the world's farmers become better agronomists, to put away their chemical crutch and transform 40 percent of Earth's surface into a bountiful food-producing, carbon-sequestering sink, instead of the largest single carbon dioxide–emitting source on the planet. We have to stop plundering our global fisheries and give wild populations time to recover, while investing in sustainable aquaculture systems that help clean up our coastal waters and put more healthy aquatic protein on our tables. We have to stop cutting down our tropical forests and start seriously investing in renewable energy to make the leap to a carbon-free economy.

If this sounds like a modern version of Condorcet's utopia, it is. Yet there are encouraging signs that a greener revolution is already under way. The organic-farming movement is thriving—as Rodale's Coach Smallwood predicted—as more and more farmers see the economic and environmental benefits and the market for organic products expands. Despite the controversy over the current generation of GMOs, some have demonstrated large environmental benefits, with little documented harm. In the hands of public researchers like Pam Ronald and Jane Langdale, the potential to use our technology to create a truly sustainable agriculture now seems within our grasp.

The best vehicle in the world in 2014, according to *Consumer Reports*, was the all-electric Tesla Model S. And while the sleek, $70,000 sedan is currently too expensive for average consumers, the number of electric, hybrid, and fuel-efficient vehicles is on the rise around the globe—just as electric-car engineer Ray Hobbs foresaw. Homes and buildings are increasingly energy-efficient as well, while the cost of solar and wind power continues to fall. In the United States the consumption of meat, particularly beef, continues to decline as the cost rises and people try to eat healthier diets lower on the food chain, while young entrepreneurs and fishmongers, like Brian O'Hanlon in Panama and Perry Rasso in Rhode Island and millions of Chinese carp/rice farmers, are helping to transform aquaculture from aquatic feedlots to efficient and sustainable fish farms that are pumping out healthy protein to a hungry world.

Researchers like Amrik Singh and his SRI farmers in India are showing the world just what simply better agronomy can produce. Young, smart, social entrepreneurs like Peter Frykman at Driptech are abandoning safe, lucrative job tracks to use their skills to create better food, better profits for poor farmers, and thus a better world. Even my alma mater, North Carolina State, once a bastion of green-revolution thinking, has established a thriving Center for Environmental Farming Systems that is striving to create "just and equitable food and farming systems that conserve natural resources, strengthen communities, improve health outcomes, and provide economic opportunities in North Carolina and beyond."

Even in Africa, peace, democracy, and growth are on the rise, while conflict is on the wane. This has fueled tremendous investment in agriculture and African economies, which are among the fastest-growing on the planet. Rwanda, once one of the most troubled nations on the continent, is now leading the charge in rapidly reducing its fertility rate through strong government support for family planning services. Perhaps more importantly, the continent's renewable energy resources—sun, wind, hydro, and geothermal—are even more abundant than its newfound fossil-fuel deposits, creating the opportunity to make the demographic transition with renewable energy instead of coal, gas, and oil.

All these pioneers who are working on the frontier of creating sustainable food systems have one thing in common: an abiding commitment to the natural world. The forester, conservationist, and farmer Aldo Leopold described this commitment 60 years ago as the "land ethic," in which humans stopped seeing themselves as conquerors of the land community but rather citizens of it, with respect for all its members. And, as Rachel Bezner Kerr found out in Malawi, healthy soils create healthy food that nurtures healthy children, mothers, and families. We all, no matter how big a city we call home, have a personal responsibility for the health of the land and waters that sustain us. Every day we make thousands of choices that affect that health—from what we drive, to what we eat, to what we buy and what we throw away. Our impact is now so large that it extends far beyond the mere terrasphere to the farthest reaches of the aquasphere and the atmosphere. If Leopold were alive, he might expand his philosophy into an Earth ethic: "A thing is right when it tends to preserve the integrity, stability, and beauty of the [Earth]. It is wrong when it tends otherwise."

Leopold knew that our estrangement from the land was the biggest roadblock to caring for its health, and we are far more isolated from it now than we were in his day. We give our children phones instead of toy shovels, iPads instead of fishing poles. It doesn't have to be that way. Shoo them out of the virtual world and let them dig for earthworms in the real one. Plant a garden and get dirty with them. Take them fishing or for a hike. Go birding in a marsh where you can steep them and yourselves in the satisfying smell of pluff mud and briny things. Take them to a farm, then to a farmer's market, and then cook a meal with them so they'll know without guessing where their food was born. Take them to a county fair and let them pet the sheep and the calves and the chickens so that they'll understand that even though we raise these animals to eat, they are part of the community too. Let them play in a rushing stream and discover the yielding yet powerful force of nature within them, as well as in you. Learn to go with the flow.

Perhaps the most telling indicator occurred in 2013, when the world's urban population finally surpassed its rural population. If the past is any indication of the future, increased urbanization should lead to better jobs, better education, lower fertility rates, an older population, and eventually democracy. But only if the land remains healthy enough to sustain our growing cities and get us to the point where the world makes the demographic transition and the global population starts to fall. If it doesn't, Malthus's positive checks of famine, war, and disease are always waiting in the wings.

THE DAY I visited Bath just happened to be a Sunday. In the midst of my hunt for Malthus's tomb, an usher informed me it was time for Choral Evensong and that all tourists would have to leave the church or stay for the service. Soft beams of afternoon sun—what the photographers at *National Geographic* call "God light"—was pouring in the windows, illuminating the scene in an ethereal glow. I chose to stay and listen. The Old Testament reading was one of my favorites from childhood, when the prophet Jonah emerges from the whale to tell the Ninevites to change their ways or be obliterated from the face of Earth. Nineveh complied and was spared.

It seems an appropriate lesson for our world.

ACKNOWLEDGMENTS

As a piece of science journalism, this book is more a synthesis of the work of others than a creation of my own, and it would not have been possible without the research of the hundreds of scientists whose names fill the notes and bibliography. I'm particularly grateful to those who took time from their busy schedules to explain their work to me in person, as well as the farmers, aquaculturists, and development workers who showed me around their fields and projects. Reviewers Charles Mann, Marion Nestle, and Paul Ehrlich were extremely gracious with their comments and suggestions and made this a much better book. Any errors in analysis or in putting the dots together are mine alone.

This book is also the culmination of more than a decade of reporting—the vast majority on assignment for *National Geographic* magazine. I'm extremely indebted to my colleagues there for sending me to the far corners of the globe to report on natural resource issues, particularly Chris Johns and Dennis Dimick, two aggies from Oregon State who have never forgotten their roots, as well as Victoria Pope, Oliver Payne, Barbara Paulsen, Rob Kunzig, Lynn Addison, Glenn Oeland, Tim Appenzeller, Peter Miller, Bob Booth, Don Belt, Carolyn White, Jennifer Reek, David Jefferies, Lisa Moore Laroe, and the late, great, John G. Mitchell, my mentor and friend. My fellow contributing writers, staff writers, former "legendeers," and photographers I've worked with remain an inspiration to me, while the magazine's incredible research staff made me a far better reporter than I ever hoped to be. I'm also forever in the debt of James G. Deane, my former editor at *Defenders* maga-zine, who was as much a lion in the field of environmental journal-ism as his famous namesake was on the silver screen.

I'd also like to thank my editors at Norton, Tom Mayer and Ryan Harrington, for having the patience of Job and the editing pencils of E. B. White, copy editor Stephanie Hiebert, who kindly and professionally cleaned up the written messes I made, and my agent Chris Parris-Lamb, who pulled a brook trout out of the stream of journalism and turned it into a whale of a tale.

Most of all, this book would never have been written without the endless support of my own farm team, Sam, Anna, and Elea-nor, whose incredible enthusiasm for all living things gives me hope for the future; and our manager, Kelly, who willingly chopped weeds out of organic cotton while eight months pregnant, sold sweet corn on the highway, grew the world's best-tasting tomatoes with manure tea, raised a menagerie of kids and animals—often single-handedly while I was on the other side of the world—and somehow still found time to help research, fact-check, and foot-note this book, giving invaluable reality checks where needed.

Yes guys, it's done.

NOTES

Introduction: The Erstwhile Agronomist

3 **"People are fighting":** "Warning over World Food Shortages," Al Jazeera, March 12, 2008, http://www.aljazeera.com/news/middleeast/2008/03/2008 525133438179651.html.

4 **The world contains more than 50,000:** T. Loftas, ed., *Dimensions of Need: An Atlas of Food and Agriculture* (Rome: Food and Agriculture Organization of the United Nations, 1995), http://www.fao.org/docrep/U8480E/U8480E00.htm.

4 **only three—wheat, rice, and corn:** F. Reifschneider, "Save Our Seeds," *Our Planet* 162, no. 2 (2005): 28–29, http://www.ourplanet.com/imgversn/162/Francisco%20Reifschneider.pdf.

5 **the fastest doubling in human history:** "International Data Base—World Population: 1950–2050," US Census Bureau, International Programs, http://www.census.gov/population/international/data/idb/worldpopgraph.php, accessed June 23, 2014; "Seven Billion and Growing: The Role of Population Policy in Achieving Sustainability," Technical Paper no. 2011/3 ([New York]: United Nations Department of Economic and Social Affairs, Population Division, [2011]), 5, http://www.un.org/esa/population/publications/technicalpapers/TP2011-3_SevenBillionandGrowing.pdf.

5 **our annual grain production rose even faster:** Lester Brown, "Feeding Everyone Well: A Status Report," chap. 7 in *Eco-Economy: Building an Economy for the Earth* (New York: Norton, 2001), http://www.earth-policy.org/books/eco/eech7_ss2.

5 **The change was so dramatic:** "Green Revolution: Curse or Blessing?," International Food Policy Research Institute, 2002, http://www.ifpri.org/sites/default/files/publications/ib11.pdf.

5 **nearly a third of the US population:** "Timeline of Farming in the U.S.: 1850–1933," *American Experience* (PBS), 1999, http://www.pbs.org/wgbh/amex/trouble/timeline.

5   **In a 2013 survey:** Jenny Hope, "The Pupils Who Think Cheese Is a Vege-
    table and Fish Fingers Come from Chickens: Study Highlights Primary
    Children's Ignorance of Food," *Daily Mail*, June 2, 2013, http://www.daily
    mail.co.uk/news/article-2334915/Pupils-think-cheese-veg-fish-fingers
    -come-chickens-Study-highlights-primary-childrens-ignorance-food-comes
    -from.html.

5   **Another survey found:** Saffron Howdon, "Cultural Cringe: Schoolchildren
    Can't See the Yoghurt for the Trees," *Sydney Morning Herald*, March 5,
    2012, http://www.smh.com.au/national/education/cultural-cringe-schoolchil
    dren-cant-see-the-yoghurt-for-the-trees-20120304-1ub55.html.

7   **When I asked a wildlife expert:** William E. Palmer and Peter T. Bromley,
    "Pesticides & Wildlife—Cotton," North Carolina Cooperative Extension
    Service AG-463-4 fact sheet, http://ipm.ncsu.edu/wildlife/cotton_wildlife
    .html.

12 **Egyptians eat more wheat:** "Egypt Grain and Feed Annual: Wheat and
    Corn Production on the Rise," GAIN Report, USDA Foreign Agricultural
    Service, Global Agricultural Information Network, April 2, 2012, http://
    gain.fas.usda.gov/Recent%20GAIN%20Publications/Grain%20and%20
    Feed%20Annual_Cairo_Egypt_4-2-2012.pdf.

12 **Bread is so important there:** "Egyptian Bread Crisis Stirs Anger," Voice of
    America, October 27, 2009, http://www.voanews.com/english/news/a-13
    -2008-03-24-voa49-66637432.html.

12 **The price of bread:** "Egyptians Riot over Bread Crisis," *Telegraph*, April 8,
    2008, http://www.telegraph.co.uk/finance/economics/2787714/Egyptians
    -riot-over-bread-crisis.html.

13 **Although the high dam at Aswan:** "International Data Base: Egypt," US
    Census Bureau, International Programs, last updated December 2013,
    http://www.census.gov/population/international/data/idb/information
    Gateway.php.

13 **Today they harvest:** Rudy Ruitenberg, Salma El Wardany, and Ola Galal,
    "Egypt's Wheat Farmers Hobbled by Fuel Shortages as Silos Run Low,"
    Bloomberg, May 9, 2013, http://www.bloomberg.com/news/2013-05-09/
    egypt-s-wheat-farmers-hobbled-by-fuel-shortages-as-silos-run-low.html.

13 **Egypt is now the world's largest importer:** "Egypt," GIEWS Country
    Briefs, Global Information and Early Warning System on Food and Agri-
    culture, last updated February 7, 2014, http://www.fao.org/giews/country
    brief/country.jsp?code=EGY.

13 **Between 2005 and 2008:** Joachim von Braun, "The Food Crisis Isn't Over,"

Nature 456, no. 11 (December 2008): 701, http://www.nature.com/nature/journal/v456/n7223/full/456701a.html.

13 **The rioters ultimately forced:** Joseph Guyler Delva and Jim Loney, "Haiti's Government Falls after Food Riots," Reuters, April 12, 2008, http://www.reuters.com/article/idUSN1228245020080413.

13 **Even in the United States:** Claudia Kalb, "Food Insecurity Rising in America," *Newsweek*, August 10, 2010, http://mag.newsweek.com/2010/08/10/food-insecurity-rising-in-america.html.

14 **By 2012 the number had doubled:** "Supplemental Nutrition Assistance Program Participation and Costs," US Department of Agriculture, last updated April 4, 2014, http://www.fns.usda.gov/pd/SNAPsummary.htm.

14 **"This is not a supply shock":** Christopher Barrett, interview by the author, September 18, 2008.

14 **The problem was so dire:** "2013 World Hunger and Poverty Facts and Statistics," World Hunger Education Service, Hunger Notes, http://www.worldhunger.org/articles/Learn/world%20hunger%20facts%202002.htm.

14 **A month from that mark:** "Hunger Statistics," World Food Programme, http://www.wfp.org/hunger/stats.

15 **roughly the size of the populations:** "Country Comparisons: Population," The World Factbook, Central Intelligence Agency, https://www.cia.gov/library/publications/the-world-factbook/rankorder/2119rank.html.

15 **As horrific as that is:** International Food Policy Research Institute, "Media Advisory," 2012, http://us1.campaign-archive2.com/?u=e1537016b431a91504702d94b&id=6ea9f7f6bf&e=5e047183c1; Alexander J. Stein and Matin Qainn, "The Human and Economic Cost of Hidden Hunger," *Food and Nutrition Bulletin* 28, no. 2 (2007), http://www.fao.org/fsnforum/sites/default/files/resources/Stein%20and%20Qaim.pdf.

15 **The world's farmers produce:** Mark Bittman, "How to Feed the World," *New York Times*, October 14, 2013, http://www.nytimes.com/2013/10/15/opinion/how-to-feed-the-world.html?pagewanted=all&_r=0.

15 **When nearly half the planet:** "Poverty Overview," World Bank, last updated April 7, 2014, http://www.worldbank.org/en/topic/poverty/overview.

16 **US diplomats defended the policy:** Joel K. Bourne Jr., "Dirt Poor," *National Geographic*, September 2008.

16 **"It has not worked":** Amy Goodman and Kim Ives, "'We Made a Devil's Bargain': Fmr. President Clinton Apologizes for Trade Policies That Destroyed Haitian Rice Farming," Democracy Now!, April 1, 2010, http://www.democracynow.org/2010/4/1/clinton_rice.

16 **Since 2000, however:** "World Grain Production, Consumption, and Stocks, 1960–2010," in "World on the Edge—Food and Agriculture Data—Crops," Earth Policy Institute, 2010, http://www.earth-policy.org; "Grain Market Report," GMR no. 404 (International Grains Council, 2010); "FAO Cereal Supply and Demand Brief," Food and Agriculture Organization of the United Nations, World Food Situation, March 7, 2014, http://www.fao .org/worldfoodsituation/csdb/en.

16 **In 2007, world grain reserves:** Joel K. Bourne Jr., "The End of Plenty," *National Geographic*, June 2009.

17 **yet world meat consumption:** Cassandra Brooks, "Consequences of Increased Global Meat Consumption on the Global Environment—Trade in Virtual Water, Energy & Nutrients," Stanford Woods Institute for the Environment, http://woods.stanford.edu/environmental-venture-projects/ consequences-increased-global-meat-consumption-global-environment.

17 **land devoted to biofuel crops:** World Bank, *World Development Report 2010: Development and Climate Change* (Washington, DC: World Bank, 2010), 147.

17 **Other agricultural experts:** Ron Meador, "Can We Feed the Planet without Trashing It? Jon Foley's 'Yes' Wins Major Award," MinnPost, March 4, 2014, http://www.minnpost.com/earth-journal/2014/03/can-we-feed-planet -without-trashing-it-jon-foleys-yes-wins-major-award.

17 **"We'll have to learn to produce":** Ashley Braun, "Purdue's Gebisa Ejeta on the Vexing Task of Feeding a Growing Population," Grist, October 14, 2010, http://grist.org/article/2010-10-13-purdues-gebisa-ejeta-on-vexing -task-feeding-growing-population.

18 **In 2010, Russia:** Dim Coumou and Stefan Rahmstorf, "A Decade of Weather Extremes," *Nature Climate Change* 2 (2012), 491–96, doi:10.1038/ nclimate1452.

18 **Back-to-back droughts:** "NCDC Releases 2012 Billion-Dollar Weather and Climate Disasters Information," National Oceanic and Atmospheric Administration, National Climatic Data Center, http://www.ncdc.noaa.gov/ news/ncdc-releases-2012-billion-dollar-weather-and-climate-disasters -information.

18 **The drought-driven food price spikes:** "Global Food Crisis Response Program: Quick Responses to Facilitate Longer-Term Solutions," World Bank, April 11, 2013, http://www.worldbank.org/en/results/2013/04/11/ global-food-crisis-response-program-results-profile.

18 **In 2013, our emissions were on track:** "Four Energy Policies Can Keep

the 2°C Climate Goal Alive," International Energy Agency, June 10, 2013, https://www.iea.org/newsroomandevents/pressreleases/2013/june/name ,38773,en.html.

18 **a broad review of climate change studies:** Rachel Warren, "The Role of Interactions in a World Implementing Adaptation and Mitigation Solutions to Climate Change," *Philosophical Transactions of the Royal Society. A: Mathematical, Physical & Engineering Sciences* 369 (2011): 217–41, doi:10.1098/rsta.2010.0271.

19 **Ice cores in Greenland:** Matthew W. Schmidt and J. E. Hertzberg, "Abrupt Climate Change during the Last Ice Age," *Nature Education Knowledge* 3, no. 10 (2011): 11, http://www.nature.com/scitable/knowledge/library/abrupt-climate-change-during-the-last-ice-24288097.

19 **The not-so-subtle warning:** R. Boyd Richerson and R. L. Bettinger, "Was Agriculture Impossible during the Pleistocene but Mandatory during the Holocene? A Climate Change Hypothesis," *American Antiquity* 66 (2001): 387–411, http://www.sscnet.ucla.edu/anthro/faculty/boyd/AgOrigins.pdf.

20 **Large swaths of the globe:** Warren, "Role of Interactions."

Chapter 1. The Curse

23 **"The superior power of population is repressed":** Thomas Robert Malthus, *An Essay on the Principle of Population*, ed. Philip Appleman, Norton critical ed. (New York: Norton 1976), 56.

23 *Homo sapiens* **migrated out of Africa:** Paul Mellars, "Why Did Modern Human Populations Disperse from Africa ca. 60,000 years ago? A New Model." *Proceedings of the National Academy of Sciences USA* 103 (2006): 9381–86, http://www.pnas.org/cgi/doi/10.1073/pnas.0510792103.

23 **Even rudimentary agriculture:** David Christian, "Contingency, Pattern and the S-Curve in Human History," World History Connected, 2009, http://worldhistoryconnected.press.illinois.edu/6.3/christian.html, accessed January 21, 2011.

24 **For the first 4 million years:** Ibid.

24 **"every species of animals":** Philip Appleman, ed., introduction to *An Essay on the Principle of Population*, by Thomas Robert Malthus, Norton critical ed. (New York: Norton, 1976), xiv.

24 **An estimated 315,000 births:** Phelim P. Boyle and Cormac Ó Gráda, "Fertility Trends, Excess Mortality, and the Great Irish Famine," *Demog-*

raphy 23 (1986): 543–62, http://EconPapers.repec.org/RePEc:spr:demogr
:v:23:y:1986:i:4:p:543-562.

25 **"I think I may fairly make":** T. R. Malthus, *An Essay on the Principle of Population* (London: J. Johnson, 1798).

25 **We now know the writer:** Donald Winch, *Malthus* (New York: Oxford University Press, 1987), 12.

27 **Rid the world of these unjust organizations:** Appleman, introduction to *An Essay*, vii–xi. Godwin's daughter, Mary Godwin Shelley, twisted that dream into a nightmare in 1818 when she created the monster of Dr. Frankenstein.

27 **As an acute observer of rural poverty:** Patricia James, ed., "Biographical Sketches," in *The Travel Diaries of Thomas Robert Malthus* (London: Cambridge University Press, 1966), 7.

27 **the bloody Reign of Terror:** Donald Winch, "Malthus, Thomas Robert," in *Encyclopedia of Population*, ed. Paul Demeny and Geoffrey McNicoll, vol. 2 (New York: Macmillan Reference USA, 2003), 619–21. Digitized version by Google Books (http://books.google.com).

28 **Ben Franklin's remarkably accurate estimates:** Conway Zirkle, "Benjamin Franklin, Thomas Malthus and the United States Census," *Isis* 48 (1957): 58–62, http://www.jstor.org/stable/226902.

29 **The Bengal Famine of 1770:** Cormac Ó Gráda, *Famine: A Short History* (Princeton, NJ: Princeton University Press, 2009), 15, http://press .princeton.edu/chapters/s8857.pdf. This number has been questioned by historians.

29 **"the whole train of common diseases":** T. R. Malthus, *An Essay on the Principle of Population, or, A View of Its Past and Present Effects on Human Happiness* (2nd ed., 1803), ed. Donald Winch, Cambridge Texts in the History of Political Thought (Cambridge: Cambridge University Press, 1992), 23.

29 **The Great Plague of London:** "The Great Plague of London, 1665," in *Contagion: Historical Views of Diseases and Epidemics*, Harvard University Library Open Collections Program, http://ocp.hul.harvard.edu/contagion/ plague.html, accessed March 29, 2014.

29 **In Malthus's day:** Robert Woods, *The Demography of Victorian England and Wales* (Cambridge: Cambridge University Press, 2000), 4. Digitized version by Google Books (http://books.google.com).

30 **"gigantic inevitable famine stalks":** Malthus, *An Essay* (Norton Critical ed.), 56.

30 **Malthus believed such cycles:** Ibid., 36.

30 **half the average at the time:** Woods, *Demography*, 5.

30 **"The natural inequality":** Malthus, *An Essay* (Norton Critical ed.), 20.

31 **"has no claim of *right*":** Patricia James, *Population Malthus* (London: Routlege & Kegan Paul, 1979), 100. This offensive passage appears only in the 1803 edition and was deleted in the 1806 edition, never to reappear, but it has been quoted against Malthus ever since.

31 **Marxists and socialists continue:** Appleman, introduction to *An Essay*, xxiv.

31 **"the greatest of all compensations":** Ibid., xxiii.

32 **Although Malthus was a lifelong advocate:** Ibid., xviii.

32 **"had better do it":** Charles Dickens, *A Christmas Carol in Prose: Being a Ghost Story of Christmas* (London: Bradbury & Evans, 1845), 14. Digitized version by Google Books (http://books.google.com).

33 **The peasant system, little changed:** Mark Overton, "Agricultural Revolution in England 1500–1850," BBC, last updated February 17, 2011, http://www.bbc.co.uk/history/british/empire_seapower/agricultural_revolution_01.shtml.

33 **"machines, for turning herbage":** Roger J. Wood and Vítezslav Orel, *Genetic Prehistory in Selective Breeding: A Prelude to Mendel* (Oxford: Oxford University Press, 2001), 38, as cited in Jonathan Harwood's book review in the *British Journal for the History of Science* 36, (2003): 239–41, http://www.jstor.org/stable/4028241.

33 **The turnips and clover were fed:** "On the Breeding, Rearing, Fattening, and General Management of Neat Cattle," *Farmer's Register*, December 1834, 393–98. Digitized version by Google Books (http://books.google.com).

33 **As a result, wheat yields grew:** Overton, "Agricultural Revolution."

34 **A full third is owned:** Tamara Cohen, "Look Who Owns Britain: A Third of the Country Still Belongs to the Aristocracy," *Mail* online, November 10, 2010, http://www.dailymail.co.uk/news/article-1328270/A-Britain-STILL-belongs-aristocracy.html.

34 **More important, the enclosure:** Overton, "Agricultural Revolution."

35 **five times more in one century:** Lester Brown, "Feeding Nine Billion," in *State of the World, 1999: A Worldwatch Institute Report on Progress toward a Sustainable Society*, ed. Linda Starke (New York: Norton, 1999), 115–32.

35 **"Malthus has been proved wrong":** Christopher Barrett, interview by the author, September 18, 2008.

36 **"Most of the people who say":** Timothy Dyson, interview by the author, October 27, 2008.

37 **When they applied Malthus's second:** Timothy Dyson, interview by the author, February 3, 2010; E. A. Wrigley and R. S. Schofield, *The Population History of England, 1541–1871: A Reconstruction* (Cambridge: Cambridge University Press, 1989), Introductory Note, xx–xxii; E. A. Wrigley, "Standing between Two Worlds: Why Malthus Is So Easily Misunderstood." Paper prepared for the Malthus Bicentenary Conference of the National Academies Forum and the National Library of Australia, September 17–18, 1998, http://pandora.nla.gov.au/pan/53598/20051026-0000/www.naf.org.au/papers.htm.

37 **Perhaps that is why:** Dyson, interview, October 27, 2008; "World Food Trends and Prospects to 2025," *Proceedings of the National Academy of Sciences USA* 96 (1999): 5929–36, doi:10.1073/pnas.96.11.5929.

Chapter 2. Famine's Lethal Lessons

39 **"A series of calamities":** M. Afzal Husain, "Minute," in Famine Inquiry Commission, *Report on Bengal* (Government of India, 1945), 195.

40 **"Not a hut was standing":** Prasanta Pramanik, *Romanthan*, quoted in "Cyclones and Floods at Contai," Contai Information Point, 2011, http://www.contai.info/cyclone1.php.

40 **Midnapur was the major:** Sir Henry Knight, *Food Administration in India, 1939–47* (Stanford, CA: Stanford University Press, 1954), 80.

40 **It had fed peasants:** Richard M. Eaton, *The Rise of Islam and the Bengal Frontier, 1204–1760* (Berkeley: University of California Press, 1993), 18, http://ark.cdlib.org/ark:/13030/ft067n99v9.

40 **"Rice is vitality":** Gyula Wojtilla, *History of Krsisastra* (Wiesbaden, Germany: Otto Harrassowitz, 2006), 16. Digitized version by Google Books (http://books.google.com).

41 **Even though more than 90 percent:** Cormac Ó Gráda, "The Ripple That Drowns? Twentieth-Century Famines in China and India as Economic History," *Economic History Review* 61 (2008): 5–37.

41 **the state still needed to import:** Omkar Goswami, "The Bengal Famine of 1943: Re-examining the Data," *Indian Economic and Social History Review* 27 (1990): 445–63, doi:10.1177/001946469002700403.

41 **Bengal's average harvest:** Cormac Ó Gráda, "Sufficiency and Sufficiency

and Sufficiency": Revisiting the Bengal Famine of 1943–44, UCD Centre for Economic Research Working Paper WP10/21 (Dublin: UCD School of Economics, University College Dublin, 2010), 15.

41 **This "boat denial policy":** Goswami, "Bengal Famine of 1943," 449; Ó Gráda, *"Sufficiency".*

41 **By some estimates:** Amartya Sen and Jean Drèze, *The Amartya Sen and Jean Drèze Omnibus: Comprising Poverty and Famines, Hunger and Public Action, India: Economic Development and Social Opportunity* (New Delhi: Oxford University Press, 1999; 10th impression, 2006), 58.

41 **In February 1943:** Ó Gráda, *"Sufficiency"*, 13.

42 **By midsummer, rice was selling:** Ibid., 18–19.

42 **Then they started selling:** Ibid., 7, 16, 27; Sen and Drèze, *Amartya Sen and Jean Drèze Omnibus*, 55.

42 **By the time a huge *aman* crop:** Timothy Dyson, interview by the author, February 3, 2010. See also Tim Dyson and Arup Maharatna, "Excess Mortality during the Bengal Famine: A Re-evaluation," *Indian Economic and Social History Review* 28, no. 3 (1991): 281–97.

43 **"But though the principle":** Donald Winch, *Malthus* (New York: Oxford University Press, 1987), 37.

43 **economic historian Cormac Ó Gráda:** Cormac Ó Gráda, *Famine: A Short History* (Princeton, NJ: Princeton University Press, 2009), 36, http://press.princeton.edu/chapters/s8857.pdf.

43 **That's 8 percent:** This figure is based on the average of world population estimates of 980 million in 1800 and 1.65 billion in 1900, from "World Population: Historical Estimates of World Population," US Census Bureau, International Programs, 2013, https://www.census.gov/population/international/data/worldpop/table_history.php.

44 **Many still accuse Stalin:** Mortality estimates range from 4 to 8 million in Ukraine and other areas of the Soviet Union. Andrea Graziosi estimates 5 million in "The Soviet 1931–1933 Famines and the Ukrainian Holodomor: Is a New Interpretation Possible, and What Would Its Consequences Be?" *Harvard Ukrainian Studies* 27 (2004–5): 97–115; while *The Oxford Handbook of the History of Communism*, edited by Stephan A. Smith (Oxford: Oxford University Press, 2014), 409, puts the range between 6 and 8 million.

44 **The famine death toll:** Stephen Devereux, "Famine in the Twentieth Century," Institute of Development Studies, Working Paper 105, January 1, 2000, http://www.ids.ac.uk/publication/famine-in-the-twentieth-century.

44   **the Chairman's disastrous attempt:** Ó Gráda, "Ripple That Drowns?" 5–6.

44   **"It was not a famine":** "Amartya Sen—Biographical," Nobel Foundation, 1998, http://www.nobelprize.org/nobel_prizes/economic-sciences/laureates/1998/sen-bio.html.

44   **Though his first love:** Christian List, "Social Choice Theory," in *Stanford Encyclopedia of Philosophy*, Winter 2013 ed., December 18, 2013, http://plato.stanford.edu/archives/win2013/entries/social-choice.

45   **He looked instead:** "Amartya Sen—Biographical."

45   **Even perceived shortages:** Amartya Sen, *Poverty and Famines: An Essay on Entitlement and Deprivation* (New York: Oxford University Press, 1981), 45–51.

45   **yet even in that case:** Steven Scalet and David Schmidtz, "Famine, Poverty, and Property Rights," in *Amartya Sen*, ed. Christopher Morris (New York: Cambridge University Press, 2010), 170–71.

46   **"the serious shortage":** Sen, *Poverty and Famines*, 53.

46   **a year in which there had been no famine:** Ibid., 58.

46   **The Bengal government's boat denial policy:** Ó Gráda, *"Sufficiency"*.

46   **"Famines are, in fact, so easy to prevent":** Amartya Sen, *Development as Freedom* (New York: Knopf, 2001), 175, cited in Thomas Plümper and Eric Neumayer, "Famine Mortality, Rational Political Inactivity, and International Food Aid," *World Development* 37 (2009): 50–61, doi:10.1016/j.worlddev.2008.05.005.

47   **For his vast body of work:** "Amartya Sen—Facts," Nobel Foundation, http://www.nobelprize.org/nobel_prizes/economic-sciences/laureates/1998/sen-facts.html, accessed April 7, 2014.

47   **The FIC estimated:** Husain, "Minute," 182.

47   **The FIC blamed the sharp fall:** Ibid., 179.

47   **Sen scoffed at this:** Sen, *Poverty and Famines*, 80.

48   **"It seems safe to conclude":** Ibid., 63, table 6.2.

48   **Instead, the British Bulldog:** Sashi Tharoor, "The Ugly Briton," *Time*, November 29, 2010; Ó Gráda, *"Sufficiency"*, 33.

48   **Husain pointed out:** Mark B. Tauger, "Entitlement, Shortage, and the 1943 Bengal Famine: Another Look," *Journal of Peasant Studies* 31 (2003): 60–63.

49   **Some evaluations of this method:** Peter Bowbrick, "The Causes of Famine: A Refutation of Professor Sen's Theory," http://bowbrick.org.

uk/Publications/The Causes of Famine 1986.pdf, cited in *A Critique of Professor Sen's Theory of Famines* (Oxford: Institute of Agricultural Economics, 1986), 111.

49   **"With statistics so hopelessly defective":** Husain, "Minute," 181.

49   **Increased wartime demand:** Ibid., 196.

49   **On the trip to his new post:** S. Y. Padmanabhan, "The Great Bengal Famine," *Annual Review of Phytopathology* 11 (1973): 11–24, doi:10.1146/annurev.py.11.090173.000303.

50   **"Though administrative failures":** Ibid.; also Mark Tauger, interview by the author, June 2011. Tauger, an expert in Soviet famines at West Virginia University who rediscovered Padmanabhan's report on brown spot disease, came to the same conclusion in "Entitlement, Shortage." Sen, while not questioning Padmanabhan's data, has challenged his and Tauger's conclusion that brown spot disease led to a major crop failure, writing in response to Tauger in the *New York Review of Books* that data from two out of two dozen research stations do not a famine make. Tauger and Sen conducted a protracted debate on this subject in the letters section of the *New York Review of Books* in the spring of 2011. While several researchers have challenged Sen's estimates of both the food supplies in 1943 and the number who died from the famine (Bowbridge, Goswami, Ó Gráda, Dyson, Maharatna, Husain, and Tauger, among others), I have found no researchers other than Sen himself who defend his numbers.

51   **Although other famines followed:** Devereaux, "Famine." Other famines on this scale include the postwar Soviet famine, China's Great Leap Forward famine, the Cambodia famine of 1979 that killed more than 1.5 million, or the North Korea famine of the late 1990s that is estimated to have killed some 3 million.

51   **The global public:** Amartya Sen and Jean Drèze, *India: Economic Development and Social Opportunity* (New Delhi: Oxford University Press, 1995), 75–76.

52   **The countries of sub-Saharan Africa:** For grain production since 1961, see FAOSTAT, Food and Agriculture Organization of the United Nations, Statistical Division, http://faostat3.fao.org/faostat-gateway/go/to/home/E; for fertility rates in China, India, Russia and sub-Saharan Africa, see "Total Fertility Rate," UNdata (searchable database from "World Population Prospects: The 2012 Revision"), http://data.un.org/Data.aspx?d=PopDiv&f=variableID:54; for fertility rates in Africa from 1950 to 2000,

see Michel M. Garenne, "Fertility Changes in Sub-Saharan Africa," DHS Comparative Reports 18, USAID, 2008, xi, http://www.measuredhs.com/pubs/pdf/CR18/CR18.pdf.

52 **Yet India continued to suffer:** Famine Inquiry Commission, cited in N. P. Nawani, "Historical Perspective of Food Management in India," in *Indian Experience on Household Food and Nutrition Security*, FAO report, August 1994, http://www.fao.org/docrep/x0172e/x0172e03.htm.

52 **"First of all, obviously":** I saw this quote in a display at the Indian Center for Agricultural Research (ICAR) Agricultural Museum, New Delhi, India, in October 2008. It can also be found in the CGIAR (Consultative Group on International Agricultural Research) newsletter *ICRISAT Happenings*, http://issuu.com/icrisat/docs/happenings_1543/8; and in M. S. Swaminathan, "From Green to an Ever-Green Revolution," in *Science and Sustainable Food Security: Selected Papers of M. S. Swaminathan* (Singapore: World Scientific, 2009), 2, doi:10.1142/9789814282116_0001.

Chapter 3. The Green Revolution: Food, Sex, and War

55 **"If you desire peace":** Norman Borlaug, "The Green Revolution, Peace, and Humanity" (Nobel lecture, December 11, 1970), http://www.nobelprize.org/nobel_prizes/peace/laureates/1970/borlaug-lecture.html.

55 **He stands in the middle:** R. Ortiz et al., "Dedication: Norman E. Borlaug, the Humanitarian Plant Scientist Who Changed the World," *Plant Breeding Reviews* 28 (2007).

56 **Within a few decades:** Lord Boyd Orr, the world-renowned nutritionist and first director of the FAO, stated in 1950 that "a lifetime of malnutrition and actual hunger is the lot of at least two-thirds of mankind." More recent research has revised that number downward to 34 percent, according to David Grigg, in Robert W. Kates, "Review of *The World Food Problem 1950–1980* by David Grigg," *Economic Geography* 63, no. 2 (1987): 183–84. Hunger stats for 2011–13 are from FAO's Hunger Portal, http://www.fao.org/hunger/en.

56 **He never forgot how a full belly:** "Interview: Norman Borlaug: Ending World Hunger," Academy of Achievement, May 12, 2008, http://www.achievement.org/autodoc/printmember/bor0int-1.

57 **A 1952 book named Stakeman:** Tim Brady, "Combating the 'Shifty Little Enemies,'" *Minnesota*, Winter 2011. Brady's article does not provide the title of the 1952 book mentioned.

57 **Land reform and rust epidemics:** Ortiz, "Dedication," 10.

58 **Rockefeller's "Mexican Agricultural Program":** John J. McKelvey Jr., *J. George Harrar 1906–1982: A Biographical Memoir*, Biographical Memoirs 57 (Washington, DC: National Academy of Sciences, 1987), 32.

59 **"he just crossed the hell out of them":** Coffman interview in the film *Freedom from Famine: The Norman Borlaug Story* ([Dayton, OH]: Mathile Institute for the Advancement of Human Nutrition, 2009).

59 **Borlaug's two locations:** Ortiz, "Dedication," 11.

60 **the rust-resistant complex of genes:** Ibid., 14.

61 **In 1965, Mexico harvested:** Ibid., 13.

62 **Thanks to Borlaug and colleagues:** "Green Revolution: Curse or Blessing?," International Good Policy Research Institute, 2002, http://www.ifpri.org/sites/default/files/publications/ib11.pdf.

62 **A rumor even started circulating:** Ortiz, "Dedication," 4.

62 **"In a restless Asia":** John Kerry King, "Rice Politics," *Foreign Affairs* 31 (1953): 453–60, http://www.jstor.org/stable/20030978.

62 **"At best the world food outlook":** Michael E. Latham, *The Right Kind of Revolution: Modernization, Development, and U.S. Foreign Policy from the Cold War to the Present* (Ithaca, NY: Cornell University Press, 2011), 112. Digitized version by Google Books (http://books.google.com).

63 **a "Manhattan Project for food":** Nick Cullather, *The Hungry World: America's Cold War Battle against Poverty in Asia* (Cambridge, MA: Harvard University Press, 2010), 162. Digitized version by Google Books (http://books.google.com).

63 **The same logic inspired:** Ibid., 161.

63 **Such government purchases:** Harry M. Cleaver Jr., "The Contradictions of the Green Revolution," *American Economic Review* 62, no. 1/2 (1972): 177–86, http://www.jstor.org/stable/1821541.

64 **"lay the basis for a permanent expansion":** "The History of Food Aid: Food for Peace Program," Alliance for Global Food Security, http://foodaid.org/resources/the-history-of-food-aid, accessed August 28, 2014.

64 **"To me that was good news":** Cleaver, "Contradictions," 2.

64 **In 1954, Mexico's undersecretary:** Ortiz, "Dedication," 16.

65 **"brown-tipped sharp-legged thing":** "Agronomy: The Rice of the Gods," *Time*, June 14, 1968. http://www.time.com/time/magazine/article/0,9171,900147,00.html.

66 **"It was an epiphany!":** Gene Hettel, "Luck Is the Residue of Design," IRRI Pioneer Interviews, *Rice Today* 5, no. 4 (2010): 10. A full transcript of the

interview with Peter Jennings can be found at https://en-gb.facebook.com/IRRI.ricenews/posts/161836633864700.

67 **"We're going to make history!":** Tom Hargrove and W. Ronnie Coffman, "Breeding History," *Rice Today* 5, no. 4 (2010): 34–38.

67 **As they approached the rice:** Hettel, "Luck Is the Residue."

67 **"If we are to win":** Hargrove and Coffman, "Breeding History."

68 **but the green revolution in rice:** Felicia Wu and William Butz, *The Future of Genetically Modified Crops: Lessons from the Green Revolution* (Santa Monica, CA: Rand Corporation, 2004), 20, http://www.rand.org/content/dam/rand/pubs/monographs/2004/RAND_MG161.pdf.

68 **The South Vietnamese government:** Latham, *Right Kind of Revolution*, 114.

69 **The United States was:** Wu and Butz, *Future of Genetically Modified Crops*, 27.

69 **India's "ship to mouth" existence:** M. S. Swaminathan, "From Green to an Ever-Green Revolution," in *Science and Sustainable Food Security: Selected Papers of M. S. Swaminathan* (Singapore: World Scientific, 2009), 2, doi:10.1142/9789814282116_0001.

69 **A few "neo-Malthusian" academics:** William Paddock and Paul Paddock, *Famine 1975! America's Decision: Who Will Survive?* (Boston: Little, Brown, 1967), 219–22.

69 **After Kennedy's assassination:** Cleaver, "Contradictions," 2 (long version).

69 **The plant exploded 15 years later:** "Rallies Held over Bhopal Disaster," BBC, December 3, 2004, http://news.bbc.co.uk/2/hi/south_asia/4064527.stm.

70 **"It was a wonderful experience":** Swaminathan, "From Green to an Ever-Green," 4.

71 **"What do you want, paper or bread?":** Norman Borlaug, in the film *Freedom from Famine*.

72 **The combination of Borlaug's wheat:** Gregg Easterbrook, "Forgotten Benefactor of Humanity," *Atlantic*, January 1997, http://www.theatlantic.com/magazine/archive/1997/01/forgotten-benefactor-of-humanity/306101.

72 **By 1991, wheat production:** Wu and Butz, *Future of Genetically Modified Crops*, 21.

72 **"These [record yields]":** William S. Gaud, "The Green Revolution: Accom-

plishments and Apprehensions" (address, Society for International Development, March 8, 1968), AgBioWorld, http://www.agbioworld.org/biotech-info/topics/borlaug/borlaug-green.html.

72 **Oddly enough, the monocultures:** "Green Revolution: Curse or Blessing?"

72 **Noted economist Jeffrey Sachs:** Jeffrey D. Sachs, *The End of Poverty: Economic Possibilities for Our Time* (New York: Penguin, 2005), 70. Digitized version by Google Books (http://books.google.com).

73 **Incredibly, all that extra food:** Easterbrook, "Forgotten Benefactor." In 1950, the world produced 692 million tons of grain on 1.7 billion acres. In 1992, it produced 1.9 billion tons on 1.73 billion acres. That amounts to a 170 percent increase in production from a 1 percent increase in land area. Also James R. Stevenson et al., "Green Revolution Research Saved an Estimated 18–27 Million Hectares from Being Brought into Agricultural Production," *PNAS* 110 (2013): 8363–8368, doi: 10.1073/pnas.1208065110.

73 **Borlaug himself is credited:** Ibid.

73 **Pesticide use tripled:** D. Tilman et al., "Forecasting Agriculturally Driven Global Environmental Change," *Science* 292 (2001): 281, doi:10.1126/science.1057544.

73 **The annual dead zone:** Mark Schleifstein, "Dead Zone the Size of Connecticut Expected along Louisiana Coast, Scientists Say," *New Orleans Times-Picayune*, June 24, 2014, http://www.nola.com/environment/index.ssf/2014/06/low_oxygen_dead_zone_the_size.html.

74 **"greed revolution":** Swaminathan, "From Green to an Ever-Green," 2.

74 **"The initiation of exploitive agriculture":** Ibid., 6.

74 **Borlaug's Nobel lecture:** Borlaug, "Green Revolution, Peace, and Humanity."

Chapter 4. The Plight of the Punjab

78 **and much of India's phenomenal:** Jeffrey D. Sachs, *The End of Poverty: Economic Possibilities for Our Time* (New York: Penguin, 2005), 180–81. Digitized version by Google Books (http://books.google.com).

79 **In 2004–5, they harvested:** N. S. Tiwana et al., *State of Environment: Punjab-2007* (Chandigarh, India: Punjab State Council for Science & Technology, 2007), xxiii, 75, http://www.npr.org/documents/2009/apr/punjab_report.pdf.

79 **This is the land of:** L. Giosan et al., "Fluvial Landscapes of the Harap-

pan Civilization," *Proceedings of the National Academy of Sciences USA* 109, no. 26 (2012): 10138–39, doi:10.1073/pnas.1112743109; Ahmad H. Dani and Vadim M. Masson, eds., *History of Civilizations of Central Asia*, vol. 1 (New Delhi: Motilal Banarsidass, 1999), 131, 287, 301. Digitized version by Google Books (http://books.google.com).

79 **a "model agricultural province":** Ian A. Talbot, "The Punjab under Colonialism: Order and Transformation in British India," *Journal of Punjab Studies* 14, no. 1 (2007): 11–27, http://www.global.ucsb.edu/punjab/14.1_Talbot.pdf.

80 **In a few years Punjabi farmers:** Vendana Shiva, interview by the author, October 2008.

81 **"Punjab state, like the rest of India":** P. S. Rangi, interview by the author, October 2008.

81 **The intensive irrigation required:** S. K. Shakya and J. P. Singh, "New Drainage Technologies for Salt-Affected Waterlogged Areas of Southwest Punjab, India," *Current Science* 99, no. 2 (2010): 204–12, http://www.currentscience.ac.in/Downloads/article_id_099_02_0204_0212_0.pdf.

82 **Punjabi farmers apply more fertilizer:** Tiwana et al., *State of Environment*, xxiv.

82 **Instead of the blistering:** "India: Basic Information," USDA Economic Research Service, May 30, 2012, http://www.ers.usda.gov/topics/international-markets-trade/countries-regions/india/basic-information.aspx#.U3-JZS_J47A.

82 **From 2001 to 2011, India added:** *Census of India 2011*, Government of India, 38–39, http://censusindia.gov.in.

82 **Though its urban middle class:** "Poverty & Equity: India," World Bank, 2014 (data for 2010), http://povertydata.worldbank.org/poverty/country/IND, accessed March 3, 2014; "India: Overview," World Food Programme, http://www.wfp.org/countries/wfp-innovating-with-india/overview, accessed March 3, 2014.

83 **Her family once grew:** Kartaro Kaur, interview by the author, October 2008.

84 **Kaur's sons were not alone:** P. Sainath, "Farm Suicides: A 12-Year Saga," *Hindu*, January 25, 2010; P. Sainath, "In 16 Years, Farm Suicides Cross a Quarter Million," *Hindu*, October 29, 2011; P. Sainath, "Some States Fight the Trend but . . . ," *Hindu*, December 5, 2011, http://www.thehindu.com/opinion/columns/sainath.

84   **"The first thing to go":** Inderjit Singh Jaijee, interview by the author, October 2008.

85   **"Cancer Express":** Praveen Donthi, "Cancer Express," *Hindustan Times*, January 16, 2010, http://www.hindustantimes.com/News-Feed/India/ Cancer-Express/Article1-498286.aspx.

85   **Many blamed the high volume:** J. Dich et al., "Pesticides and Cancer," *Cancer Causes & Control* 8, no. 3 (1997): 420–43.

85   **Organophosphates are neurotoxins:** Q. Li, "New Mechanism of Organophosphorus Pesticide-Induced Immunotoxicity," *Journal of Nippon Medical School* 74, no. 2 (2007): 92–105; "Facts about Sarin," Centers for Disease Control and Prevention, last updated May 20, 2013, http://www.bt.cdc .gov/agent/sarin/basics/facts.asp.

86   **The link between pesticides:** Tiwana et al., *State of Environment*, 113–16.

87   **The results were striking:** The study found 125 cases per 100,000 people in the south, versus 72 cases per 100,000 in the north. Annual cancer deaths were similar, with 52 per 100,000 among cotton farmers versus 30 per 100,000   among the northern wheat and rice growers. These rates are still less than half those in developed countries, where older, fatter, less active populations smoke more, drink more alcohol, and consume diets higher in fats and refined carbohydrates—Western lifestyles that help give them the highest cancer rates in the world. See "Global Cancer Rates Could Increase by 50% to 15 million by 2020," World Health Organization, April 3, 2003, http://www.who.int/mediacentre/news/releases/2003/ pr27/en/#.

87   **Residents in the cotton belt:** J. S. Thakur et al., "Epidemiological Study of High Cancer among Rural Agricultural Community of Punjab in Northern India," *International Journal of Environmental Research and Public Health* 5, no. 5 (2008): 399–407.

88   **"Over the last four years":** Jagsir Singh, interview by the author, October 2008. The others quoted in this passage—Amarjid Kaur, Tej Kaur, Billu Singh, Sima Singh, Jagdeve Singh, Bhola Singh, and Gurdeep Singh— were all interviewed during the same week in October 2008.

88   **each sum about the total average:** K. Vatta, B. R. Garg, and M. S. Sidhu, "Rural Employment and Income: The Inter-household Variations in Punjab," *Agricultural Economics Research Review* 21, no. 2 (2008): 201–10.

90   **Several recent studies, however:** P. R. Jonnalagadda et al., "Genotoxicity in Agricultural Farmers from Guntur District of South India—A Case

Study," *Human & Experimental Toxicology* 31, no. 7 (2012): 741–47, http://het.sagepub.com/content/31/7/741.abstract; J. A. Bhalli et al., "DNA Damage in Pakistani Agricultural Workers Exposed to Mixture of Pesticides," *Environmental and Molecular Mutagenesis* 50, no. 1 (2009): 37–45, doi:10.1002/em.20435; R. Naravaneni and K. Jamil, "Determination of AChE Levels and Genotoxic Effects in Farmers Occupationally Exposed to Pesticides," *Human & Experimental Toxicology* 26, no. 9 (2007): 723–31, doi:10.1177/0960327107083450; D. S. Rupa, P. P. Reddy, and O. S. Reddi, "Clastogenic Effect of Pesticides in Peripheral Lymphocytes of Cotton-Field Workers," *Mutation Research*, 261, no. 3 (1991): 177–80.

90 **In North Carolina, no one can buy:** "NC Pesticide Safety Education Program," North Carolina State University, http://ipm.ncsu.edu/pesticide safety, accessed April 24, 2014. For required safety equipment, see "Personal Protective Equipment (PPE)," Southern Region Pesticide Safety Education Center, http://ipm.ncsu.edu/srpsec/ppe.htm.

90 **In India, it's a different story:** A 2010 survey of farmworkers in Maharashtra found conditions similar to those that the Bhuttiwala villagers had described on my visit there, with 88 percent of workers saying they used no precautions when handling or spraying pesticides. Dhanraj A. Patil and Ravasaheb J. Katti, "Modern Agriculture, Pesticides and Human Health: A Case of Agricultural Labourers in Western Maharashtra," *Journal of Rural Development*, 31, no. 3 (2012): 305–18, http://www.indiaenviron mentportal.org.in/files/file/agriculture%20&%20pesticides.pdf.

91 **When I asked officials:** Manjit Singh Kang (former vice-chancellor, Punjab Agricultural University), interview by the author, October 2008.

91 **"Nobody has suggested":** Jagsir Singh, interview, October 2008.

92 **"The green revolution has brought":** Jarnail Singh, interview by the author, October 2008.

94 **some 30,000 Punjabis had died:** Shiva, interview, October 2008. For background on Punjab violence, see "Dead Silence: The Legacy of Human Rights Abuses in Punjab," Human Rights Watch, May 1, 1994, http://www.hrw.org/reports/1994/05/01/dead-silence-legacy-abuses-punjab; "India: Break the Cycle of Impunity and Torture in Punjab," Amnesty International, January 20, 2003, http://www.amnesty.org/en/library/info/ASA20/002/2003.

95 **Researchers have documented:** M. J. Moechnig et al., "Empirical Corn Yield Loss Estimation from Common Lambsquarters (*Chenopodium*

album) and Giant Foxtail (*Setaria faberi*) in Mixed Communities," *Weed Science* 51, no. 3 (2003): 386–93, http://www.jstor.org/stable/4046674.

95 **In 2004, researchers at Johns Hopkins:** V. Singh and K. P. West Jr., "Vitamin A Deficiency and Xerophthalmia among School-Aged Children in Southeastern Asia," *European Journal of Clinical Nutrition* 58 (2004): 1342–49, doi:10.1038/sj.ejcn.1601973.

96 **"Sometime in the mid-1990s":** G. S. Kalkat, interview by the author, October 2008.

Chapter 5. China: Landraces and Lamborghinis

99 **"When I make enough money":** Sun Haipeng, interview by the author, October 2008.

101 **The Chinese were among the first:** "Chinese Pigs 'Direct Descendants' of First Domesticated Breeds," ScienceDaily, April 20, 2010, http://www .sciencedaily.com/releases/2010/04/100419150947.htm, reporting on G. Larson et al., "Patterns of East Asian Pig Domestication, Migration, and Turnover Revealed by Modern and Ancient DNA," *Proceedings of the National Academy of Sciences USA* 107, no. 17 (April 27, 2010): 7686–91, doi:10 .1073/pnas.0912264107.

101 **When the distraught man grabbed:** "Yorkshire," Oklahoma State University, Breeds of Livestock, http://www.ansi.okstate.edu/breeds/swine/ yorkshire, accessed January 5, 2012.

102 **Few outside China:** David Pilling, "Why Beijing Must End Inflation—or Else," *Financial Times*, May 18, 2011, http://www.ft.com/cms/s/0/1bf6dc04 -817c-11e0-9c83-00144feabdc0.html#axzz1iWnnp5lA.

102 **from about 20 kilograms a year:** Chen Yaosheng (Sun Yat-sen University), interview by the author, October 19, 2008; James Hansen, Fred Gale, "China in the Next Decade: Rising Meat Demand and Growing Imports of Feed," *Amber Waves*, April 7, 2014, USDA Economic Research Service, http://www.ers.usda.gov/amber-waves/2014-april/china-in-the-next-decade -rising-meat-demand-and-growing-imports-of-feed.aspx#.VKHHyUAZ8Y accessed on 12/29/2014.

102 **more than 40 in 2014:** "China Livestock and Products Semi-annual Report 2011," The Pig Site, March 24, 2011, http://www.thepigsite.com/processing/ articles/1247/china-livestock-and-products-semiannual-report-2011.

102 **the People's Republic consumes half:** *China's Agricultural Trade: Com-*

petitive Conditions and Effects on U.S. Exports, USITC Publication 4219 (Washington, DC: US International Trade Commission, 2011), xvii, http://www.usitc.gov/publications/332/pub4219.pdf.

103   **Times are good:** All quotes from merchants in the Construction New Village Market come from interviews conducted by the author in October 2008.

104   **the modern superpigs:** "Danish Landrace," Oklahoma State University, Breeds of Livestock, http://www.ansi.okstate.edu/breeds/swine/danishland race, accessed January 5, 2012.

105   **these still produce 70 percent:** Matt Phillips, "A Solution to China's Food Price Inflation Problem," National Pork Producers Council, November 19, 2010, http://www.nppc.org/wp-content/uploads/numbering.pdf.

105   **Infected sows often abort:** "Porcine Reproductive & Respiratory Syndrome (PRRS)," The Pig Site, http://www.thepigsite.com/diseaseinfo/97/porcine-reproductive-respiratory-syndrome-prrs, accessed January 9, 2012.

105   **though Chinese scientists:** "Emergency Vaccination Alleviates HP-PRRS Virus Infection," Pig Progress, July 30, 2013, http://www.pigprogress.net/Health-Diseases/Research/2013/7/Emergency-vaccination-alleviates-HP-PRRS-virus-infection-1322484W.

105   **Over the next few years:** David Barboza, "Virus Spreading Alarm and Pig Disease in China," *New York Times*, August 16, 2007.

105   **but a 2011 report:** *China's Agricultural Trade*, 6–10.

105   **Even without such pandemics:** Barboza, "Virus Spreading Alarm."

105   **China's sows are now:** Phillips, "Solution to China's Food Price," 4.

106   **such factory farms have unleashed:** Robert Paarlberg, "The Changing Politics of CAFOs," Farm Foundation AgChallenge2050, March 1, 2013, http://www.agchallenge2050.org/farm-and-food-policy/2013/02/the-changing-politics-of-cafos.

106   **and nearly a thousand:** Joel K. Bourne Jr., "The End of Plenty," *National Geographic*, June 2009.

107   **"There's still room for":** Qingzhang Lu, interview by the author, October 2008.

107   **Soy imports have grown:** Karl Plume, "Chinese Importers Sign Deals to Buy $2.8 Billion of US Soybeans," Reuters, September 16, 2013, http://www.reuters.com/article/2013/09/16/usa-soybeans-china-idUSL2N0H81L620130916. For 2000 figures, see Francis Tuan, Cheng Fang, and Zhi Cao, "China's Soybean Imports Expected to Grow Despite Short-Term Disruptions," USDA Economic Research Service, October 2004, 3,

http://www.ers.usda.gov/publications/ocs-oil-crops-outlook/ocs04j01.aspx#.U_PXK2PDX9g.

107 **The nation now consumes:** "Oilseeds: World Markets and Trade," USDA Foreign Agricultural Service, May 2014, http://www.fas.usda.gov/data/oil seeds-world-markets-and-trade.

107 **Another seismic shift:** Hansen, "China in the Next Decade."

108 **"In 2008, as long as you kept a pig":** Shen Guang Rong, interview by the author, October 2008.

111 **"sprouted from the Machong banana fields":** "Cargill Protein Feed (Dongguan) Ltd., Cargill-President (Dongguan) Feed Protein Technology Co., Ltd.: About Us," Cargill, http://www.cargill.com.cn/en/locations/dongguan/index.jsp, accessed January 17, 2012.

111 **a 35 percent increase:** "Financial Information: Fiscal 2011," Cargill, http://www.cargill.com/company/financial/index.jsp.

112 **The "paper empress":** "Rich CPPCC Member Kicks Up Storm," *People's Daily* Online, March 10, 2008, http://english.people.com.cn/90001/90776/90882/6369627.html.

113 **17 percent of its domestic production:** Peng Gong, "China Needs No Foreign Help to Feed Itself," *Nature* 474, no. 7 (2011), doi:10.1038/474007a.

113 **more than the average annual grain harvest of Russia:** In 2012, Russia grew 88 million tons of grain: "Russian Federation Grain and Feed Annual for 2012," GAIN Report RS1219, USDA Foreign Agricultural Service, Global Agricultural Information Network, March 30, 2012.

113 **It's helped to drive a rapid expansion:** "Geographers Predict Increasing Rate of Amazon Deforestation," ScienceDaily, July 15, 2011, http://www.sciencedaily.com/releases/2011/07/110714120722.htm. See also P. Vitousek et al., "Globalization, Trade, and the Environment: The Case of Brazil," Freeman Spogli Institute for International Studies at Stanford University, http://fsi.stanford.edu/research/2275, accessed January 19, 2012.

114 **"I don't remember the year":** Yaosheng, interview, October 19, 2008.

114 **In 2009 the Chinese government:** "National Policy Responses to Cereal Price Spikes during 2007–2011," in Agricultural Market Information System, "Enhancing Market Transparency," *Food Outlook: Global Market Analysis*, November 2011, 23.

115 **The Chinese government estimates:** Jialin Zhang, "China's Slow-Motion Land Reform: Tentative Steps and Halting Progress," *Policy Review*, February 1, 2010.

115 **More than 3 million hectares:** David Stanway, "China Says Over 3 Mil-

lion Hectares of Land Too Polluted to Farm," Reuters, December 30, 2013, http://www.reuters.com/article/2013/12/30/china-environment-farm land-idUSL3N0K90OY20131230.

115 **As much as 60 percent:** Andrew Jacobs, "China Says It Curbed Spill of Toxic Metal in River," *New York Times*, January 30, 2012, http://www .nytimes.com/2012/01/31/world/asia/china-says-it-curbed-spill-of-toxic -metal-in-river.html?_r=0.

115 **Today, half of China's rivers:** Ibid.

116 **Each year, China consumes:** Marlys Miller, "China Pork Imports to Set a Record," PorkNetwork, last updated November 29, 2011, http://www .porknetwork.com/pork-news/Chinese-pork-imports-to-reach-record -high-in-2011-134670638.html.

116 **Placed end to end:** Dividing 50 million metric tons by 26 metric tons per 38-foot refrigerated container gives 1,923,076 containers, which, end to end, would stretch 73,076,923 feet or, at 5,280 feet per mile, 13,840 miles.

116 **A recently declassified report:** *Global Trends 2030: Alternate Worlds* (National Intelligence Council, 2012), http://www.dni.gov/nic/globaltrends; Mathew Burrows, "Climate Change Impact on National Security" (Climate Lecture 17, University of Copenhagen, September 2009).

118 **One prominent China watcher:** Thomas Robert Malthus, *An Essay on the Principle of Population*, ed. Philip Appleman, Norton critical ed. (New York: Norton, 1976), 109.

Chapter 6. Food, Fuel, and Profit

121 **"Alcohol can be manufactured":** Alexander Graham Bell, "Prizes for the Inventor: Some of the Problems Awaiting Solution," *National Geographic* 31 (February 1917), 133.

122 **So they added 2 percent gasoline:** Lindsay Goldwert, "Indy 500 Goes from Gas to Green," CBS News, May 25, 2007, http://www.cbsnews.com/ 2100-18563_162-2853518.html.

122 **Such homegrown fuels would bolster:** Twenty-eight percent of greenhouse gas emissions in the United States come from our transportation. "Sources of Greenhouse Gas Emissions," US Environmental Protection Agency, http://www.epa.gov/climatechange/ghgemissions/sources.html, April 4, 2012.

123 **Though the ethanol industry claims:** "American Ethanol Enhances

Partnership with Richard Childress Racing and Driver Austin Dillon in 2014," American Ethanol, February 20, 2014, http://americanethanolracing .com/news/american-ethanol-enhances-partnership-with-richard-childress -racing-and-driver-austin-dillon-in-2014.

123 **If the stills are coal fired:** Michael Wang, May Hu, and Hong Ho, "Life-Cycle Energy and Greenhouse Gas Emission Impacts of Different Corn Ethanol Plant Types," *Environmental Research Letters* 2, no. 2 (2007): 12, doi:10.1088/1748-9326/2/2/024001.

123 **For newer plants fired by natural gas:** EPA, *Renewable Fuel Standard Program (RFS2) Regulatory Impact Analysis*, EPA-420-R-10-006 (Environmental Protection Agency, 2010), 471, http://www.epa.gov/otaq/renewable fuels/420r10006.pdf.

123 **though even that number is optimistic:** Sasha Lyutse, "EPA's RFS Accounting Shows Corn Ethanol Today Is Worse than Gasoline," *Switchboard* (blog), Natural Resources Defense Council, July 20, 2010, http:// switchboard.nrdc.org/blogs/slyutse/as_i_discussed_here_last.html.

124 **Biodiesel's greenhouse gas emissions:** Dev Shrestha, "Environmental Life Cycle Analysis of Biodiesel," eXtension, January 31, 2014, http://www .extension.org/pages/27999/environmental-life-cycle-analysis-of-biodiesel.

124 **A palm-oil plantation:** Damian Carrington, "Leaked Data: Palm Biodiesel as Dirty as Fuel from Tar Sands," *Environment* (blog), *Guardian*, January 27, 2012, http://www.guardian.co.uk/environment/damian-car rington-blog/2012/jan/27/biofuels-biodiesel-ethanol-palm-oil.

124 **We, the SUV-loving people:** US Energy Information Administration, http://www.eia.gov.

124 **About 100 million older cars:** "E15 Approved for Use in 2001 and Newer Vehicles," US Department of Energy, Alternative Fuels Data Center, February 11, 2011, http://www.afdc.energy.gov/bulletins/technology_bulletin_ 1210.html.

125 **"blend wall":** National Research Council, *Renewable Fuel Standard: Potential Economic and Environmental Effects of U.S. Biofuel Policy* (Washington, DC: National Academies Press, 2011), 383, http://www.nap.edu/openbook .php?record_id=13105&page=383.

125 **So why are we now planting:** Dina Cappiello and Matt Apuzzo, "The Secret Environmental Cost of US Ethanol Policy," AP News, November 12, 2013, http://bigstory.ap.org/article/secret-dirty-cost-obamas-green -power-push-1.

125 **"It is a crime against humanity":** "UN Independent Rights Expert Calls for Five-Year Freeze on Biofuel Production," UN News Centre, October 26, 2007, http://www.un.org/apps/news/story.asp?NewsID=24434&#.U_v-lmPDX9g.

125 **and the Egyptians were worshipping Osiris:** David J. Hanson, "History of Alcohol and Drinking around the World," adapted from *Preventing Alcohol Abuse: Alcohol, Culture and Control* (Westport, CT: Praeger, 1995), http://www2.potsdam.edu/alcohol/Controversies/1114796842.html#.VBmB6efJ47A.

125 **When New England inventor Samuel Morey:** Bill Kovarik, "Henry Ford, Charles Kettering and the Fuel of the Future," *Automotive History Review*, no. 32 (Spring 1998): 7–27, http://www.environmentalhistory.org/billkovarik/about-bk/research/henry-ford-charles-kettering-and-the-fuel-of-the-future.

126 **"In . . . 10 to 20 years":** Ibid.

127 **"gasohol":** "Gasohol: A Technical Memorandum," US Congress, Office of Technology Assessment, September 1979, http://books.google.com/books/about/Gasohol.html?id=iNRTAAAAMAAJ.

127 **Andreas was a poor Mennonite farm kid:** Ronald Henkoff and Sara Hammes, "Oh, How the Money Grows at ADM," *Fortune*, October 8, 1990, http://archive.fortune.com/magazines/fortune/fortune_archive/1990/10/08/74164/index.htm.

128 **A report by the Libertarian Cato Institute:** James Bovard, *Archer Daniels Midland: A Case Study in Corporate Welfare*, Cato Policy Analysis 241 (Cato Institute, 1995), http://www.cato.org/pubs/pas/pa-241.html.

128 **That number had doubled to $20 billion:** Robert Pear, "After Three Decades, Tax Credit for Ethanol Expires," *New York Times*, January 1, 2012, http://www.nytimes.com/2012/01/02/business/energy-environment/after-three-decades-federal-tax-credit-for-ethanol-expires.html?_r=0.

128 **From the 1920s to the 1970s:** Magda Lovei, *Phasing Out Lead from Gasoline: Worldwide Experience and Policy Implications*, World Bank Technical Paper 397 (Washington, DC: World Bank, 1998), http://siteresources.worldbank.org/INTURBANTRANSPORT/Resources/b09phasing.pdf; Kovarik, "Henry Ford."

128 **The oil and petrochemical industries:** "MTBE in Drinking Water," Environmental Working Group, October 22, 2003, http://www.ewg.org/research/mtbe-drinking-water.

129 **Any improvement in air pollution:** EPA, *Achieving Clean Air and Clean Water: Report of the Blue Ribbon Panel on Oxygenates in Gasoline*, EPA-420-R-99-021 (Environmental Protection Agency, 1999), http://www.epa.gov/otaq/consumer/fuels/oxypanel/r99021.pdf.

129 **Oil company engineers learned:** "Partial Settlement Requires Oil Companies to Pay Substantial Settlement and Treat Wells for MTBE over the Next 30 Years," Baron & Budd, May 7, 2008, http://baronandbudd.com/areas-of-practice/water-contamination/mtbe-settlement-press-kit/mtbe-settlement-press-release.

129 **Even though the chemical causes cancer in rats:** Patty Toccalino, "Human-Health Effects of MTBE: A Literature Summary," USGS National Water Quality Assessment Program, January 9, 2013, http://sd.water.usgs.gov/nawqa/vocns/mtbe_hh_summary.html.

130 **But despite billions:** Brad Plumer, "The U.S. May Be Hitting Its Ethanol Limit. So EPA Wants to Relax Its Biofuels Goals," *Washington Post*, November 15, 2013, http://www.washingtonpost.com/blogs/wonkblog/wp/2013/11/15/the-u-s-is-hitting-its-ethanol-limit-so-the-epa-will-weaken-its-biofuels-rules.

130 **The return on investment:** David Coltrain, "Economic Issues with Ethanol" (paper presented at the Risk and Profit Conference, Kansas State University, Manhattan, KS, August 16–17, 2001).

131 **Several EU countries:** Donald Mitchell, *A Note on Rising Food Prices*, Policy Research Working Paper 4682 (World Bank, Development Prospects Group, 2008), 9–10, https://openknowledge.worldbank.org/bitstream/handle/10986/6820/WP4682.pdf.

131 **In all, more than 40 countries:** Nomura, *The Coming Surge in Food Prices*, Global Economics and Strategy (Nomura, 2010), http://www.nomura.com/europe/resources/pdf/080910.pdf.

131 **Though palm oil is still:** R. Kongsager and A. Reenberg, *Contemporary Land-Use Transitions: The Global Oil Palm Expansion*, GLP Report 4 (Copenhagen: Global Land Project International Project Office, 2012), 17, http://www.globallandproject.org/arquivos/Kongsager,_R_and_Reenberg_A_%282012%29_Contemporary_land_use_transitions_The_global_oil_palm.pdf.

132 **The resulting smaller soybean crop:** Mitchell, *Note*, 10.

133 **With spiraling food prices:** David Goldman, "Food Price Spike: Is Ethanol to Blame?," *CNN Money*, June 27, 2008; *Testimony of Edward P. Lazear*

Chairman, Council of Economic Advisers before the Senate Foreign Relations Committee Hearing on "Responding to the Global Food Crisis," May 14, 2008, http://georgewbush-whitehouse.archives.gov/cea/lazear20080514.html.

133   **The International Monetary Fund (IMF) estimated:** Desmond Butler, "U.S. Disputes IMF on Food Prices," *USA Today*, May 5, 2008, http://usatoday30.usatoday.com/news/washington/2008-05-14-3613606601_x.htm.

133   **The IFPRI report went even further:** Mitchell, *Note*, 17.

134   **"The results of this analysis":** Bruce A. Babcock, "Updated Assessment of the Drought's Impacts on Crop Prices and Biofuels Production," CARD Policy Brief 12-BP8, Center for Agricultural and Rural Development, Iowa State University, August 2012, http://www.card.iastate.edu/publications/synopsis.aspx?id=1169; Hibah Yousuf, "Corn Prices Rally to New Record High," *CNN Money*, The Buzz, August 9, 2012, http://buzz.money.cnn.com/2012/08/09/corn-prices-record.

134   **It's a place kids leave in droves:** Cheyenne Wells, "The Great Plains Drain," *Economist*, January 17, 2008, http://www.economist.com/node/10534077.

134   **"This is the first year":** Roger Harders, interview by the author, March 12, 2007.

135   **"Combine sales are up 50 percent":** Gary Rasmussen, interview by the author, March 12, 2007.

135   **"I think the price will remain high":** Duane Anderson, interview by the author, March 12, 2007.

136   **Since 2006, US farmers have plowed up:** "Conservation Programs: Statistics," USDA Farm Service Agency, September 9, 2014, http://www.fsa.usda.gov/FSA/webapp?area=home&subject=copr&topic=rns-css.

136   **Nitrogen levels in the two rivers:** Cappiello and Apuzzo, "Secret Environmental Cost."

136   **From 2003 to 2008:** Sergio Schlesinger, *Sugar Cane and Land Use Change in Brazil: Biofuel Crops, Indirect Land Use Change and Emissions*, Briefing (Brussels: Friends of the Earth Europe, 2010), 3, http://www.foe.co.uk/sites/default/files/downloads/sugar_cane_and_land_use_ch.pdf.

137   **From 2005 to 2008:** S. R. Loarie et al., "Direct Impacts on Local Climate of Sugar Cane Expansion in Brazil," *Nature Climate Change* 1 (2011) 105–9, doi:10.1038/nclimate1067.

137   **Such land-use change:** Daniel deB Richter Jr. and R. A. Houghton, "Gross CO_2 Fluxes from Land-Use Change: Implications for Reducing Global

Emissions and Increasing Sinks," *Carbon Management* 2, no. 1 (2011): 41–47; F. N. Tubiello et al., "Agriculture, Forestry and Other Land Use Emissions by Sources and Removals by Sinks: 1990–2100 Analysis," FAO Statistics Division, Working Paper Series, ESS/14-02, March 2014, p. 20, http://www.fao.org/docrep/019/i3671e/i3671e.pdf, accessed on 12/15/14.

137 **Though the rate of deforestation:** Maria Luiza Rabello and Stephen Nielsen, "Brazil Rewrites Amazon Protections in Bid to Cut Carbon Emission," Bloomberg, December 7, 2011, http://www.bloomberg.com/news/2011-12-07/brazil-rewrites-amazon-protections-in-bid-to-cut-carbon-emission.html.

137 **The conversion resulted in massive releases:** "Oil Palm Plantations: Threat and Opportunities for Tropical Ecosystems," United Nations Environmental Programme, UNEP Global Environmental Alert Service (GEAS), December 2011, http://www.unep.org/pdf/Dec_11_Palm_Plantations.pdf.

137 **When Shell merged:** Elzio Barreto and Inae Riveras, "Shell Bets on Ethanol in $21 Billion Deal with Brazil's Cosan," Reuters, February 1, 2010, http://www.reuters.com/article/2010/02/01/us-cosan-shell-idUSTRE6101TW20100201.

138 **the Cosan group:** "Factbox: Mergers, Takeovers in Brazil's Ethanol Industry," Reuters, February 1, 2010, http://www.reuters.com/article/2010/02/01/us-cosan-shell-factbox-idUSTRE6102GM20100201.

138 **"a river of ethanol flowing":** Carlos Vinicius Xavier, Fábio T. Pitta, and Maria Luisa Mendonça, *A Monopoly in Ethanol Production in Brazil: The Cosan-Shell Merger* (Milieudefensie [Friends of the Earth Netherlands] and the Transnational Institute, 2011), 9, http://www.social.org.br/ethanol_monopoly_brazil.pdf.

138 **In 2011, armed military and paramilitary:** Eitan Haddok, "Biofuels Land Grab: Guatemala's Farmers Lose Plots and Prosperity to 'Energy Independence,'" *Scientific American*, January 13, 2012, http://www.scientificamerican.com/article.cfm?id=biofuels-land-grab-guatemala.

139 **In the report, the FAO:** *Price Volatility in Food and Agricultural Markets: Policy Responses*, Policy Report Including Contributions by FAO, IFAD, IMF, OECD, UNCTAD, WFP, the World Bank, WTO, IFPRI, and the UN HLTF, May 3, 2011, http://www.ictsd.org/downloads/2011/05/final g20report.pdf.

139 **In the six months between:** Peter Wahl, *Food Speculation the Main Factor in the Price Bubble of 2008*, Briefing Paper (Berlin: World Economy,

Ecology and Development, 2009), http://www2.weed-online.org/uploads/weed_food_speculation.pdf.

139 **One study showed that 70 percent:** Steve Suppan, "Dodd-Frank Position Limits on Commodity Contracts: Round 2," *Think Forward* (blog), Institute for Agriculture and Trade Policy, November 14, 2013, http://www.iatp.org/blog/201311/dodd-frank-position-limits-on-commodity-contracts-round-2.

140 **The same year, world wheat prices doubled:** Grace Livingstone, "The Real Hunger Games: How Banks Gamble on Food Prices—and the Poor Lose Out," *Independent*, April 1, 2012, http://www.independent.co.uk/news/world/politics/the-real-hunger-games-how-banks-gamble-on-food-prices—and-the-poor-lose-out-7606263.html#.

141 **The paper's authors called for:** Marco Lagi et al., "The Food Crises: A Quantitative Model of Food Prices Including Speculators and Ethanol Conversion," New England Complex Systems Institute, September 21, 2011, http://necsi.edu/research/social/food_prices.pdf.

141 **"There are hundreds of billions":** *Testimony of Michael W. Masters, Managing Member/Portfolio Manager, Masters Capital Management, LLC, Before the Committee on Homeland Security and Governmental Affairs, United States Senate,* May 20, 2008, 8, http://hsgac.senate.gov/public/_files/052008 Masters.pdf.

142 **Even though the federal Commodity Futures:** "History of the CFTC," US Commodity Futures Trading Commission, http://www.cftc.gov/About/HistoryoftheCFTC/history_precftc, accessed September 12, 2014.

142 **Four years after Dodd-Frank was passed:** Suppan, "Dodd-Frank Position Limits."

142 **Uncannily, the researchers had submitted:** Annia Ciezadlo, "Let Them Eat Bread: How Food Subsidies Prevent (and Provoke) Revolutions in the Middle East," *Foreign Affairs*, March 23, 2011, http://www.foreignaffairs.com/articles/67672/annia-ciezadlo/let-them-eat-bread.

Chapter 7. The Gauntlet

145 **"When acre has been added to acre":** Donald Winch, *Malthus* (New York: Oxford University Press, 1987), 17.

145 **"the greatest hunger fighter of our time":** M. S. Swaminathan, "Memorial Address" (Norman Borlaug memorial service, Texas A&M University, October 6, 2009, http://borlaug.tamu.edu/files/2010/10/Borlaug-Memorial

-Swaminathan-Remarks.pdf; Scott Kilman and Roger Thurow, "Father of 'Green Revolution' Dies," *Wall Street Journal*, September 13, 2009, http://online.wsj.com/articles/SB125281643150406425.

146 **Even wheat rust, the scourge:** Ronnie Coffman (Chair, Borlaug Global Rust Initiative, Cornell University), interview by the author, May 2, 2012; C. Saintenac et al., "Identification of Wheat Gene *Sr35* That Confers Resistance to Ug99 Stem Rust Race Group," *Science* 341, no. 6147 (August 16, 2013): 783–86, doi:10.1126/science.1239022.

147 **It's no coincidence:** Clemens Breisinger, Olivier Ecker, and Perrihan Al-Riffai, "Economics of the Arab Awakening: From Revolution to Transformation and Food Security," IFPRI Policy Brief 18 (Washington, DC: International Food Policy Research Institute, 2011), http://www.ifpri.org/sites/default/files/publications/bp018.pdf.

148 **Agriculture is, by far:** Jonathan A. Foley, "Can We Feed the World and Sustain the Planet?" *Scientific American*, November 2011, 60–65, http://www.scientificamerican.com/article/can-we-feed-the-world.

149 **Another 4.1 billion hectares:** FAO, "Sustainability Dimensions," in *FAO Statistical Yearbook 2012: Africa Food and Agriculture*, part 4 (Accra, Ghana: Food and Agriculture Organization of the United Nations, 2013), 184, http://www.fao.org/docrep/018/i3137e/i3137e00.htm.

149 **As a result, 90 percent of the increase:** "How to Feed the World in 2050," High-Level Expert Forum, Food and Agriculture Organization of the United Nations, Rome, October 12–13, 2009, http://www.fao.org/wsfs/forum2050/wsfs-forum/en.

149 **Already the amount of agricultural land:** Felicia Wu and William P. Butz, *The Future of Genetically Modified Crops: Lessons from the Green Revolution* (Santa Monica, CA: Rand Corporation, 2004), 31, http://www.rand.org/pubs/monographs/MG161.html.

149 **That's just over a third of an acre:** "Highlights of Annual 2013 Characteristics of New Housing," US Census Bureau, https://www.census.gov/construction/chars/highlights.html, accessed November 26, 2013.

150 **Meanwhile, 8 percent:** Nomura, *The Coming Surge in Food Prices*, Global Economics and Strategy (Nomura, 2010), http://www.nomura.com/europe/resources/pdf/080910.pdf.

150 **"The rich, soft soil":** David R. Montgomery, *Dirt: The Erosion of Civilizations* (Berkeley: University of California Press, 2007), 51. Digitized version by Google Books (http://books.google.com).

150   **Over the last few decades:** David R. Montgomery, "Soil Erosion and Agricultural Sustainability," *Proceedings of the National Academy of Sciences USA* 104, no. 33 (2007): 13268–72, doi:10.1073/pnas.0611508104.

150   **By some estimates:** "Unprecedented Pressures on Farmland—with 30 Million Hectares Lost Annually—Poses 'Direct Threat to the Right to Food of Rural Populations,' Third Committee Told," GA/SHC/3985, United Nations General Assembly, October 21, 2010, http://www.un.org/News/Press/docs/2010/gashc3985.doc.htm.

151   **"Globally, a return to per capita production":** Chris C. Funk and Molly E. Brown, "Declining Global Per Capita Agricultural Production and Warming Oceans Threaten Food Security," *Food Security* 1 (2009): 271–89, doi:10.1007/s12571-009-0026-y.

152   **But yields have not continued to grow:** Ibid.

152   **"We got seduced by how easy":** Peter R. Jennings, "Luck Is the Residue of Design," in *The IRRI Pioneer Interviews*, conducted by Gene Hettel (*Rice Today*, 2007), http://books.irri.org/Pioneer_Interviews.pdf.

153   **the latter amount having tripled:** UNESCO, *Managing Water under Uncertainty and Risk*, United Nations World Water Development Report 4 (Paris: United Nations Educational, Scientific and Cultural Organization, 2012), http://unesdoc.unesco.org/images/0021/002156/215644e.pdf.

153   **According to the World Bank:** Nomura, *Coming Surge*, 17.

153   **Over the next few decades:** "OECD Environmental Outlook to 2050: The Consequences of Inaction," Organisation for Economic Co-operation and Development, 2012, http://www.oecd.org/environment/oecdenvironmentaloutlookto2050theconsequencesofinaction.htm.

153   **The United Nations World Water Development Report:** Ben Block, "U.N. Raises 'Low' Population Projection for 2050," Eye on Earth, Worldwatch Institute, http://www.worldwatch.org/node/6038, accessed June 14, 2012.

153   **In early 2012, a report:** "Global Water Security," Intelligence Community Assessment, ICA 2012-08, February 2, 2012, 1, http://fas.org/irp/nic/water.pdf.

153   **The latter is an aquatic lifeline:** Andrew Quinn, "U.S. Intelligence Sees Global Water Conflict Risks Rising," Reuters, March 22, 2012, http://www.reuters.com/article/2012/03/22/us-climate-water-idUSBRE82L0PR20120322.

153   **for more than 40 million people:** "Amu Darya Basin Network," East-

West Institute, http://amudaryabasin.net/content/amu-darya-river-basin, accessed September 11, 2014.

153 **In fact, the rate of groundwater depletion:** Richard A. Kerr, "India's Groundwater Disappearing at Alarming Rate," Science Now, August 10, 2009, http://news.sciencemag.org/sciencenow/2009/08/10-01.html.

154 **In all 12 estimates of world population:** "Understanding and Using Population Projections," Population Reference Bureau, December 2001, http://www.prb.org/Publications/PolicyBriefs/UnderstandingandUsing PopulationProjections.aspx.

154 **The good news is:** "2011 World Population Data Sheet," Population Reference Bureau, July 2011, http://www.prb.org/Publications/Datasheets/2011/ world-population-data-sheet/data-sheet.aspx.

155 **In 2004, UN demographers predicted:** United Nations Population Division, *World Population to 2300* (New York: UN Department of Economic and Social Affairs, Population Division, 2004), 1, http://www.un.org/esa/ population/publications/longrange2/WorldPop2300final.pdf.

155 **In 2015, the medium-population projection:** United Nations Population Division, *World Population Prospects: The 2015 Revision: Key Findings and Advance Tables* (New York: UN Department of Economic and Social Affairs, Population Division, 2015), "Summary and Key Findings," 5, http:// esa.un.org/unpd/wpp/Publications/Files/Key_Findings_WPP_2015.pdf.

155 **Population Action International:** Timothy Dyson, interview by the author, February 23, 2010.

155 **Its studies show:** "7 Billion and Counting: What Global Population Growth Means for People and the Planet," Policy & Issue Brief, Population Action International, July 2011, http://populationaction.org/wp-content/ uploads/2011/12/7bn2011.pdf.

156 **Married men and women:** Joel E. Cohen, "Seven Billion," *New York Times*, October 23, 2011, http://www.nytimes.com/2011/10/24/opinion/seven -billion.html?pagewanted=all.

156 **According to the 2014 estimate:** Intergovernmental Panel on Climate Change, *Climate Change 2014: Synthesis Report, Contribution of Working Groups I, II and III to the Fifth Assessment Report of the Intergovernmental Panel on Climate Change*, ed. Core Writing Team, R.K. Pachauri, and L.A. Meyer (Geneva, Switzerland: IPCC, 2015), https://www.ipcc.ch/report/ar5/syr/.

157 **When they subjected the crops:** Justin Gillis, "A Warming Planet Struggles to Feed Itself," *New York Times*, June 4, 2011.

157   **"One of the things that we're starting to believe":** Ibid.

157   **Chronically malnourished Bangladesh:** World Bank, *World Develop-ment Report 2010: Development and Climate Change* (Washington, DC: World Bank, 2010), 6.

157   **According to the IPCC:** R. T. Watson et al., eds., "Tropical Asia," chap. 11 in *The Regional Impacts of Climate Change: An Assessment of Vulnera-bility* (Cambridge: Cambridge University Press for the Intergovernmental Panel on Climate Change, 1998), http://www.grida.no/publications/other/ipcc_sr/?src=/climate/ipcc/regional/index.htm.

158   **At our current rate:** Justin Gillis, "U.N. Climate Panel Endorses Ceil-ing on Global Emissions," *New York Times*, September 27, 2013, http://www.nytimes.com/2013/09/28/science/global-climate-change-report.html?pagewanted=all.

158   **No less a luminary:** National Intelligence Council, *Global Trends 2030: Alternative Worlds* (National Intelligence Council, 2012), i, http://www.dni.gov/index.php/about/organization/global-trends-2030.

159   **Laozi blames humanity's unnatural desires:** Chad Hansen, "Daoism," in *The Stanford Encyclopedia of Philosophy*, Fall 2013 ed., ed. Edward N. Zalta, http://plato.stanford.edu/archives/fall2013/entries/daoism.

159   **"lack acting and yet lack 'don't act'":** Ibid.

160   **"What we have to start looking at":** Ray Hobbs, interview by the author, March 28, 2007.

162   **"goad of necessity":** T. R. Malthus, *An Essay on the Principle of Population, or, A View of Its Past and Present Effects on Human Happiness*, ed. Donald Winch, Cambridge Texts in the History of Political Thought (Cambridge: Cambridge University Press, 1992), xix.

Chapter 8. The Blue Revolution

165   **"With Earth's burgeoning human population":** Quoted by Michael Rubino in "Offshore Aquaculture: Building on Policy, Technology & Research" (presentation to the Aquaculture America Meeting, February 2006), http://www.lib.noaa.gov/retiredsites/docaqua/presentations/aa_off shorepanel_files/rubino_aa_06.pdf.

166   **"I love to just sit":** Brian O'Hanlon, interview by the author, May 10, 2011.

166   **Opening just 2 percent of the oceans:** "Volumes of the World's Oceans

from ETOPO1," ETOPO1 Global Relief Model, NOAA, http://ngdc.noaa
.gov/mgg/global/etopo1_ocean_volumes.html, accessed August 8, 2012.

167 **The world now farms more fish than beef:** Joel K. Bourne Jr., "How to
Farm a Better Fish," *National Geographic*, June 2014.

167 **Nearly half of all the fish and shellfish:** FAO, *The State of Fisheries and
Aquaculture*, 2014: 20, 24, http://www.fao.org/3/a-i3720e.pdf.

167 **Population growth, income growth:** R. Waite et al., *Improving Produc-
tivity and Environmental Performance of Aquaculture: Creating a Sustainable
Food Future, Installment Five*, World Resources Report (World Resources
Institute, 2014), http://www.wri.org/publication/improving-aquaculture.

168 **"Within the next fifty years":** Peter F. Drucker, "Beyond the Information Revolution," *Atlantic*, October 1, 1999, http://www.theatlantic.com/
magazine/archive/1999/10/beyond-the-information-revolution/304658.

168 **Carvings on Egyptian tombs:** Bernardo Basurco and Alessandro Lova-
telli, "The Aquaculture Situation in the Mediterranean Sea—Predictions
for the Future" (paper presented at the International Conference on the
Sustainable Development of the Mediterranean and Black Sea Environment (IASON), May 28–June 1, 2003), http://hdl.handle.net/1834/543.

168 **On the other side of the planet:** Graydon "Buddy" Keala, *Loko I'a: A
Manual on Hawaiian Fishpond Restoration and Management* (Honolulu:
University of Hawai'I Press, 2007), 4–5, http://www.ctahr.hawaii.edu/oc/
freepubs/pdf/Loko%20I%27a%20Full%20Publication.pdf.

169 **Today Hawaii imports 85–90 percent:** Hawaii Office of Planning,
Department of Business, Economic Development & Tourism, *Increased
Food Security and Food Self-Sufficiency Strategy* (Honolulu: Office of Plan-
ning, 2012), http://files.hawaii.gov/dbedt/op/spb/INCREASED_FOOD_
SECURITY_AND_FOOD_SELF_SUFFICIENCY_STRATEGY.pdf.

169 **By the Three Kingdoms period:** Wenhua Li, ed., *Agro-ecological Farming Systems in China*, Man and the Biosphere Series 26 (Paris: UNESCO,
2001), 192–95. Digitized version by Google Books (http://books.google
.com).

169 **The system was so simple:** Ibid.

170 **Las Vegas alone consumes:** Jenny Slafkosky, "Sustaining the Shrimp
Supply," *Contra Costa Times*, June 25, 2008, http://www.contracostatimes
.com/living/ci_9685058.

170 **The area used for seaweed aquaculture:** Matt Walker, "Origin of Giant

Bloom Discovered," *BBC Earth News*, May 6, 2009, http://news.bbc.co.uk/earth/hi/earth_news/newsid_8026000/8026847.stm.

170 **To keep pens of densely packed fish:** David Barboza, "In China, Farming Fish in Toxic Waters," *New York Times*, December 15, 2007, http://www.nytimes.com/2007/12/15/world/asia/15fish.html?pagewanted=all.

170 **In 2006 and 2007:** D. C. Love et al., "Veterinary Drug Residues in Seafood Inspected by the European Union, United States, Canada, and Japan from 2000 to 2009," *Environmental Science & Technology* 45, no. 17 (2011): 7232–40; *Seafood Safety: FDA Needs to Improve Oversight of Imported Seafood and Better Leverage Limited Resources*, Report to Congressional Requesters, GAO-11-286 (US Government Accountability Office, 2011), http://www.gao.gov/assets/320/317734.pdf.

171 **"China preferred polyculture":** Li Sifa, interview by the author, April 15, 2011.

171 **Fish farms in Europe and Latin America:** FAO, *The State of Fisheries and Aquaculture*, 2012: 8–9, http://www.fao.org/docrep/016/i2727e/i2727e.pdf; A. Alvial et al., *The Recovery of the Chilean Salmon Industry: The ISA Crisis and Its Consequences and Lessons* (Puerto Montt, Chile: Global Aquaculture Alliance, 2012), http://www.gaalliance.org/cmsAdmin/uploads/GAA_ISA-Report.pdf.

173 **Scientists have yet to detect:** Daniel Bennetti, interview by the author, May 11, 2011.

173 **Pacific sardine populations:** Juan P. Zwolinski and David A. Demer, "A Cold Oceanographic Regime with High Exploitation Rates in the Northeast Pacific Forecasts a Collapse of the Sardine Stock," *Proceedings of the National Academy of Sciences USA* 109 (2012): 4175–80; "Climate Variability & Marine Fisheries: Collapse of Anchovy Fisheries and the Expansion of Sardines in Upwelling Regions," PEEL Climate & Marine Fisheries, http://www.pfeg.noaa.gov/research/climatemarine/cmffish/cmffishery4.html, accessed August 10, 2012.

174 **Yet so hot is the market:** Quirin Schiermeier, "Ecologists Fear Antarctic Krill Crisis," *Nature* 467, no. 15 (2010), doi:10.1038/467015a.

174 **The ratio of pounds of forage fish:** Rosamond L. Naylor, interview by the author, April 7, 2011; R. L. Naylor et al., "Feeding Aquaculture in an Era of Finite Resources," *Proceedings of the National Academy of Sciences USA* 106 (2009): 15103–10, doi:10.1073/pnas.0905235106.

175 **"My long-term vision":** O'Hanlon, interview, May 10, 2011.

176 **"The whole concept of moving aquaculture":** Stephen Cross, interview by the author, May 12, 2011.

177 **Even China is going back:** Sena S. De Silva and F. Brian Davy, eds., *Success Stories in Asian Aquaculture* (New York: Springer, 2009), 20. Digitized version by Google Books (http://books.google.com).

178 **"I'd get to the farmer's market":** Perry Rasso, interview by the author, March 1, 2011.

179 **"Net pens are a total goat rodeo":** Bill Martin, interview by author, January 31, 2011.

Chapter 9. Back in the USSR

183 **"When I am dead, bury me":** Taras H. Shevchenko, "My Testament," Schevchenko Poetry, Taras Schevchenko Museum, December 25, 1845 (trans. John Weir, 1961), http://www.infoukes.com/shevchenkomuseum/poetry.htm#link4.

184 **It lost almost 2 million hectares:** Veljko Mikelic, *Housing, Land, and Property in Crimea* (Nairobi: United Nations Human Settlements Programme, 2007), 15. Digitized version by Google Books (http://books.google.com).

184 **Buried beneath nearly 70 percent:** Mark Rackevych, "Agriculture: Room for Profitable Growth in the Fields of Black Earth," *Financial Times*, September 17, 2012; E. Eckmeier et al., "Pedogenesis of Chernozems in Central Europe: A Review," *Geoderma* 139 (2007): 288–99, http://www.geo.unizh.ch/~mschmidt/downloads/Eckmeier_Geoderma_2007.pdf.

185 **The country suffered:** *Achieving Ukraine's Agricultural Potential: Stimulating Agricultural Growth and Improving Rural Life* (Washington, DC: OECD and World Bank, 2004), xvii, http://www.oecd.org/tad/agricultural-policies/34031855.pdf.

185 **Ukraine is more famous for:** "Frequently Asked Chernobyl Questions," International Atomic Energy Agency, http://www.iaea.org/newscenter/features/chernobyl-15/cherno-faq.shtml, accessed October 17, 2012; *Integrating Environment into Agriculture and Forestry Progress and Prospects in Eastern Europe and Central Asia*, vol. 2, *Ukraine: Country Review* (Washington, DC: World Bank, 2007), 1, http://www.worldbank.org/eca/pubs/envint/Volume%20II/English/Review%20UKR-final.pdf.

186 **Take 12 million off:** World Bank, *Competitive Agriculture or State Control:*

Ukraine's Response to the Global Food Crisis (World Bank, Europe and Central Asia Region Sustainable Development Unit, 2008), 9, http://site resources.worldbank.org/INTUKRAINE/Resources/WorldFoodCrisis andRoleofUkraine.pdf.

186   **Ukraine alone has 3 million:** "Ukraine: Grain Production Prospects for 2008–2009," USDA Foreign Agricultural Service, Commodity Intelligence Report, May 15, 2008, http://www.pecad.fas.usda.gov/highlights/2008/05/ukr_15may2008/ukr_15may2008.htm; "Set #8: Mineral Soils Conditioned by a Steppic Climate," in "Lecture Notes on the Major Soils of the World," FAO Corporate Document Repository, http://www.fao.org/docrep/003/Y1899E/y1899e11.htm, accessed October 5, 2012.

186   **Yet because of the export restrictions:** "Cereal Offenders: Curbing Food Exports to Feed Hungry Mouths Is a Recipe for Trouble," *Economist*, May 27, 2008, http://www.economist.com/node/10926502.

187   **"They thought that":** All quotes from Justin Bruch in this chapter come from interviews by the author on September 12–14, 2012.

187   **Instead, Morgan Stanley lost:** Alan Katz and Peter Robison, "Morgan Stanley Bet the Farm in Ukraine before Taxpayers' Bailout," Bloomberg, October 4, 2011, http://www.bloomberg.com/news/2011-10-04/morgan-stanley-bet-the-farm-in-ukraine-before-fed-bailout-by-u-s-taxpayer.html.

187   **Swedish investment group Alpcot Agro:** Agrokultura, *Annual Report 2011* (May 18, 2012) and *Annual Report 2013* (April 24, 2014), http://www.agrokultura.com/financial%20reports.aspx.

190   **In 2011 only a third:** Serhiy Moroz, "Rural Households in Ukraine: Current State and Tendencies," *Economics of Agriculture* 60 (2013): 565–84, http://purl.umn.edu/158256.

190   **With 36 percent of rural families below:** Pavlo Rozenko (Razumkov Center), quoted in "Employment Does Not Protect a Person from Poverty in Ukraine," *Kyiv Post*, November 16, 2011; *Farm Reference Handbook for Ukraine* (USAID, 2005), 8, http://www.lol.org.ua/eng/docs/Farm%20Handbook%20ENG.pdf.

190   **That year 83 percent:** *Achieving Ukraine's Agricultural Potential*, 44–46.

191   **Together they explained:** Quotes from Vladimir Bubnov, Andrey Portrylo, and Jimmy Zimmerhanzel in the passage that follows come from interviews by the author on September 12, 2012.

194   **"The former organization of labour":** John Dixon and Aidan Gulliver, *Farming Systems and Poverty: Improving Farmers' Livelihoods in a Changing*

World, ed. Malcolm Hall (Rome: FAO and World Bank, 2001), http://www
.fao.org/3/a-ac349e.pdf.

194 **In 2013, Ukraine ranked 144th:** "Corruption Perception Index 2013,"
Transparency International, http://www.transparency.org/cpi2013/results,
accessed April 12, 2014.

194 **Anger over persistent government corruption:** "Ukraine's Crisis: A
Tale of Two Countries," *Economist*, February 24, 2014, http://www.econo
mist.com/blogs/easternapproaches/2014/02/ukraines-crisis.

198 **Several tests of Ukrainian:** Yuliya Dubinyik, "Ukraine: Agricultural Biotechnology Annual," GAIN Report UP1222, USDA Foreign Agricultural
Service, Global Agricultural Information Network, July 3, 2012, http://
gain.fas.usda.gov/Recent%20GAIN%20Publications/Agricultural%20Bio
technology%20Annual_Kiev_Ukraine_7-3-2012.pdf.

199 **It takes little fertilizer:** E. S. Oplinger et al., "Buckwheat," Alternative
Field Crops Manual, November 1989, https://www.hort.purdue.edu/new
crop/afcm/buckwheat.html.

199 **The country now has:** *Organic Agriculture—A Step towards the Green
Economy in the Eastern Europe, Caucasus and Central Asia Region—Case
Studies from Armenia, Moldova and Ukraine*, Green Economy (United Nations
Environment Programme, 2011), 7, http://www.unep.ch/etb/publications/
Green%20Economy/Organic%20Agriculture%20-%20EECCA%20study
%20on%20Green%20Economy/UNEP%20Organic_EECCA%20coun
try%20study.pdf; Oane Visser, "Land Deal by Foreign Investors in Russia,
Ukraine, & Kazakhstan" (presentation to the Global Forum for Food and
Agriculture, Berlin, January 19, 2012).

199 **Though most of the organic crops:** "Ukraine: Parliament Approves Law
on Organic Farming," Research Institute of Organic Agriculture (FiBL),
April 26, 2011, http://www.fibl.org/en/service-en/news-archive/news/article/
ukraine-parliament-approves-law-on-organic-farming.html; Swiss Agency
for Development and Cooperation, http://www.eda.admin.ch/eda/en/home/
dfa/orgcha/sectio/sdc.html.

200 **Much of Ukraine's farmland already suffers:** *Ukraine: Country Review*, 1.

200 **By adopting a nationwide program:** "Agricultural Pollution in Ukraine
(Azov-Black Sea Region)," World Bank, http://www.ais.unwater.org/ais/aiscm/
getprojectdoc.php?docid=85, accessed November 30, 2013.

Chapter 10. The Blooming Desert

204 **Two-thirds of the nation's:** "Imperial County Agriculture: Feeding the World," Imperial County Farm Bureau, http://www.icfb.net/countyag.html, accessed November 9, 2012.

204 **about 3 million acre-feet to be precise:** "About IID Water," Imperial Irrigation District, http://www.iid.com/index.aspx?page=112, accessed November 9, 2012.

204 **where the Imperial Dam diverts 20 percent:** Felicity Barringer, "Empty Fields Fill Urban Basins and Farmers' Pockets," *New York Times*, October 23, 2011, http://www.nytimes.com/2011/10/24/science/earth/24water.html?pagewanted=all&_r=0.

205 **The All-American is part of 3,000 miles:** "Agriculture Customers," Imperial Irrigation District, http://www.iid.com/index.aspx?page=151, accessed November 20, 2012.

205 **"This is one of the few places":** Lloyd Allen, interview by the author, February 2004.

206 **circling the globe 180 times:** *Agriculture at a Crossroads: Global Summary for Decision Makers* (International Assessment of Agricultural Knowledge, Science and Technology for Development, 2009), 8, http://www.unep.org/dewa/agassessment/reports/IAASTD/EN/Agriculture%20at%20a%20Crossroads_Global%20Summary%20for%20Decision%20Makers%20%28English%29.pdf.

206 **In the United States, agriculture uses:** "Irrigation and Water Use: Overview," USDA Economic Research Service, June 7, 2013, http://www.ers.usda.gov/topics/farm-practices-management/irrigation-water-use.aspx.

206 **We've stored so much water:** Peter H. Gleick, "Dam It, Don't Dam It, Undam It: America's Hydropower Future," *Huffington Post*, August 6, 2012, http://www.huffingtonpost.com/peter-h-gleick/americas-hydropower-future_b_1749182.html.

206 **Aside from beating back drought:** Gary Wolff and Peter H. Gleick, "The Soft Path for Water," chap. 1 in *The World's Water 2002–2003* (Washington, DC: Island Press, 2002), 2, 13, http://pacinst.org/publication/the-worlds-water-2002-2003.

206 **Up to 15 million salmon:** Bill M. Bakke, "Chronology of Salmon Declines in the Columbia River, 1779 to the Present Based on the Historical Record," Native Fish Society, http://www.nativefishsociety.org/conservation/

documents/CHRONCR-NWSALMONDECLINE3-12-09.pdf, accessed November 19, 2012.

206 **There are now more than 450 dams:** John Harrison, "Dams: History and Purpose," Northwest Power & Conservation Council, October 31, 2008, https://www.nwcouncil.org/history/DamsHistory.

207 **Today, despite spending:** J. Lichatowich, L. Mobrand, and L. Lestelle, "Depletion and Extinction of Pacific Salmon (*Oncorhynchus* spp.): A Different Perspective," *ICES Journal of Marine Science* 56, no. 4 (1999): 467–72; Robert T. Lackey, "Pacific Northwest Salmon: Forecasting Their Status in 2100," *Reviews in Fisheries Science* 11 (2003): 35–88.

207 **Nearly all of the stocks:** Bakke, "Chronology of Salmon," 23.

207 **In fact, all over the globe:** Doug Chadwick, "Silent Streams," *National Geographic*, April 2010.

207 **"I'm sure people can survive":** Marc Reisner, "The Fight for Reclamation," *High Country News*, March 20, 1995, http://www.hcn.org/issues/31/874.

207 **Even though only 20 percent:** Hugh Turral, Jacob Burke, and Jean-Marc Faurès, *Climate Change, Water and Food Security*, FAO Water Reports 36 (Rome: Food and Agriculture Organization of the United Nations, 2011), 23, http://www.fao.org/docrep/014/i2096e/i2096e.pdf.

208 **Since 1950, when:** David Molden, ed., *Water for Food, Water for Life: A Comprehensive Assessment of Water Management in Agriculture* (London: Earthscan, 2007), 7, http://www.iwmi.cgiar.org/assessment/Publications/books.htm.

208 **In the Imperial Valley:** *Vegetative Water Use in California, 1974*, Bulletin 113-3 (Sacramento: California Department of Water Resources, April 1975), http://www.water.ca.gov/waterdatalibrary/docs/historic/Bulletins/Bulletin_113/Bulletin_113-3__1975.pdf.

208 **Meltwater from the Himalaya:** Turral, Burke, and Faurès, *Climate Change*, xix.

209 **Already, peak runoff:** Ibid., 76.

209 **But farmers with senior water rights:** J. A. Vano et al., "Climate Change Impacts on Water Management and Irrigated Agriculture in the Yakima River Basin, Washington, USA," *Climatic Change* 102 (2010): 287–317.

209 **According to the IPCC:** "Climate Change 2007: Working Group II: Impacts, Adaptation, and Vulnerability, Executive Summary," *IPCC Fourth Assessment Report: Climate Change 2007*, http://www.ipcc.ch/publications _and_data/ar4/wg2/en/ch3s3-es.html, accessed December 16, 2014.

209 **A recent study by the FAO:** Turral, Burke, and Faurès, *Climate Change*, xvii.

209 **The math is simple:** Peter H. Gleick, interview by the author, November 13, 2012.

210 *Comprehensive Assessment:* Molden, *Water for Food*, 14.

210 **Critics contend that the sheer weight:** Michael Wines, "China Admits Problems with Three Gorges Dam," *New York Times*, May 19, 2011, http://www.nytimes.com/2011/05/20/world/asia/20gorges.html.

210 **"monument to corruption":** Wolff and Gleick, "Soft Path," 13.

210 **In 2011, two program managers:** "Second Former U.S. Army Corps of Engineers Manager Pleads Guilty in Alleged $30 Million Bribery and Kickback Scheme," FBI, Washington Field Office, May 17, 2012, http://www.fbi.gov/washingtondc/press-releases/2012/second-former-u.s.-army-corps-of-engineers-manager-pleads-guilty-in-alleged-30-million-bribery-and-kickback-scheme.

210 **One group of Alaskan entrepreneurs:** Peter H. Gleick, "Zombie Water Projects (Just When You Thought They Were Really Dead . . .)," Forbes .com, December 7, 2011, http://www.forbes.com/sites/petergleick/2011/12/07/zombie-water-projects-just-when-you-thought-they-were-really-dead; Gleick, interview, November 13, 2012.

211 **"If we could ever competitively":** John F. Kennedy, "119—The President's News Conference," American Presidency Project, April 12, 1961, question 13, http://www.presidency.ucsb.edu/ws/?pid=8055.

211 **It takes roughly 2,000 tons of water:** M. M. Mekonnen and A. Y. Hoekstra, "A Global and High-Resolution Assessment of the Green, Blue and Grey Water Footprint of Wheat," *Hydrology and Earth System Sciences Discussions* 7 (2010): 2499–2542.

211 **Though the cost of desalination:** Jeffrey Marlow, "The Pursuit of Cost-Effective Desalination," *Green* (blog), *New York Times*, September 21, 2009, http://green.blogs.nytimes.com/2009/09/21/on-the-pursuit-of-cost-effective-desalination.

211 **Agriculture, with its high-use:** Turral, Burke, and Faurès, *Climate Change*, 28.

213 **They even get better nutrient uptake:** "Laser Leveling: Resource Conservation through Laser Leveling," Department of Soil & Water Conservation, Punjab, http://dswcpunjab.gov.in/contents/Laser_Leveling.htm, accessed November 26, 2012; Swarleen Kaur, "Laser Land Leveller a Rev-

elation for Farm Sector in Punjab," *Financial Express*, September 14, 2009, http://www.financialexpress.com/news/laser-land-leveller-a-revelation-for-farm-sector-in-punjab/516588.

213 **"Farmers everywhere are vey innovative species":** Aditi Mukherji, interview by the author, November 14, 2012.

214 **"I had hardly any concept":** Peter Frykman, interview by the author, October 5, 2012.

215 **Farmers in China are even using it:** Forrest Laws, "Fertigation Can Help Boost Fertilizer Efficiency," *Southwest Farm Press*, April 21, 2009, http://southwestfarmpress.com/management/fertigation-can-help-boost-fertilizer-efficiency.

216 **Australian orchard growers:** "Water Management: Technologies That Save and Grow," Factsheet 4, Food and Agriculture Organization of the United Nations, http://www.fao.org/ag/save-and-grow/pdfs/factsheets/en/SG-water.pdf, accessed November 26, 2012.

217 **And despite legitimate concerns:** Molden, *Water for Food*, 3.

218 **"India has made billions of dollars":** Mukherji, interview, November 14, 2012.

219 **"This Salton Sea thing":** Allen, interview, February 2004.

Chapter 11. Magic Seeds: Feeding Shareholders or the World?

223 **"In my dream I see green":** Norman Borlaug, "The Green Revolution, Peace, and Humanity" (Nobel lecture, December 11, 1970), http://www.nobelprize.org/nobel_prizes/peace/laureates/1970/borlaug-lecture.html.

223 **Zimbabwe's President Robert Mugabe:** "Mugabe Orders White Farmers to Leave," BBC News World Edition, August 12, 2002, http://news.bbc.co.uk/2/hi/africa/2187453.stm.

224 **Many resorted to eating:** Noah Zerbe, "Feeding the Famine? American Food Aid and the GMO Debate in Southern Africa," *Food Policy* 29 (2004): 593–608.

225 **As a result, these countries halted:** Henri E. Cauvin, "Between Famine and Politics, Zambians Starve," *New York Times*, August 30, 2002.

225 **One British activist:** "British Fears Guide African Food Policies," *Washington Times*, January 13, 2003.

225 **Even Andrew Natsios:** Zerbe, "Feeding the Famine?"

226 **researchers at UC Davis:** Matthew B. Hufford et al., "Comparative Pop-

ulation Genomics of Maize Domestication and Improvement," *Nature Genetics* 44 (2012): 808–11.

228 **By 2008 the four largest seed companies:** Philip H. Howard, "Visualizing Consolidation in the Global Seed Industry: 1996–2008," *Sustainability* 1 (2009): 1266–87.

228 **The first significant agricultural product:** Jason McLure, "Genetically Modified Food," *CQ Researcher*, August 31, 2012, 717–39; I. R. Dohoo et al., "A Meta-analysis Review of the Effects of Recombinant Bovine Somatotropin," *Canadian Journal of Veterinary Research* 67 (2003): 252–64.

228 **At the time, the nation did not need more milk:** "Dairy Termination Program: An Estimate of Its Impact and Cost-Effectiveness," US General Accounting Office, Report to Congressional Requesters, GAO/RCED-89-96, July 6, 1989, http://www.gao.gov/assets/220/211399.pdf.

228 **Canada, the European Union:** McLure, "Genetically Modified Food."

229 **In the United States, doping:** "Steroid Hormone Implants Used for Growth in Food-Producing Animals," US Food and Drug Administration, last updated July 28, 2014, http://www.fda.gov/animalveterinary/safety health/productsafetyinformation/ucm055436.htm.

229 **By 2002, about 22 percent:** Andrew Pollack, "Which Cows Do You Trust?" *New York Times*, October 7, 2006.

229 **By 2010, less than 10 percent:** Mateusz Perkowski, "Dairymen Reject rBST Largely on Economic Grounds," *Capital Press*, December 10, 2013.

230 **Now, with the new soybeans:** David R. Huggins and John P. Reganold, "No-Till: The Quiet Revolution," *Scientific American*, July 2008, 70–77.

230 **as did the use of glyphosate:** W. A. Battaglin et al., "Glyphosate and Its Degradation Product AMPA Occur Frequently and Widely in U.S. Soils, Surface Water, Groundwater, and Precipitation," *Journal of the American Water Resources Association (JAWRA)* 50, no. 2 (2014): 275–90, doi:10.1111/jawr.12159.

230 **Rachel Carson even proposed it:** Rachel Carson, *Silent Spring*, 40th anniv. ed. from Mariner Books (New York: Houghton Mifflin, 2002), 290.

231 **By 2012 the amount of US farmland in no-till:** Frank Lessiter, "No-Till Advances Conservation Efforts," NoTillFarmer.com, November 29, 2012, http://www.no-tillfarmer.com/pages/Spre/No-Till-Advances-Conservation-Efforts-December-5,-2012.php.

231 **Glyphosate, the active ingredient in Roundup:** "Consumer Factsheet

on: Glyphosate," US Environmental Protection Agency, http://www.epa
.gov/ogwdw/pdfs/factsheets/soc/glyphosa.pdf, accessed July 1, 2014.

231 **This was a vast improvement:** Elizabeth A. Warnemuende et al., "Effects
of Tilling No-Till Soil on Losses of Atrazine and Glyphosate to Runoff Water under Variable Intensity Simulated Rainfall," *Soil and Tillage
Research* 95 (2007): 19–26, doi:10.1016/j.still.2006.09.001.

231 **The latter, still widely used:** Tyrone B. Hayes et al., "Demasculinization
and Feminization of Male Gonads by Atrazine: Consistent Effects across
Vertebrate Classes," *Journal of Steroid Biochemistry and Molecular Biology*
127 (2011): 64–73, doi:10.1016/j.jsbmb.2011.03.015.

231 **It is a ubiquitous contaminant:** Paul D. Winchester, Jordan Huskins,
and Jun Ying, "Agrichemicals in Surface Water and Birth Defects in the
United States," *Acta Paediatrica* 98 (2009): 664–69, doi:10.1111/j.1651
-2227.2008.01207.x.

231 **Bt crops in 2011 alone:** Graham Brookes and Peter Barfoot, "Key Environmental Impacts of Global Genetically Modified (GM) Crop Use 1996–
2011," *GM Crops and Food* 4, no. 2 (2013): 109–19.

232 **"no verifiable untoward toxic":** FAO, "Health and Environmental
Impacts of Transgenic Crops," chap. 5 in *The State of Food and Agriculture:
2003–2004* (Rome: Food and Agricultural Organization of the United
Nations, 2004), http://www.fao.org/docrep/006/y5160e/y5160e10.htm#P0_0.

232 **Most of those safety claims:** M. Antoniou et al., "Teratogenic Effects of Glyphosate-Based Herbicides: Divergence of Regulatory Decisions from Scientific Evidence," *Journal of Environmental & Analytical Toxicology* S4 (2012): 006.

232 **Early studies did ring alarm bells:** Laurie Flynn and Michael Gillard,
"Pro-GM Food Scientist 'Threatened' Editor,'" *Guardian*, October 31,
1999; Peta Firth, "Leaving a Bad Taste," *Scientific American*, May 1999.

233 **given that the EPA has allowed:** Carey Gillam, "Angry Mothers Meet
U.S. EPA over Concerns with Roundup Herbicide," Reuters, May 27,
2014, http://www.reuters.com/article/2014/05/27/us-monsanto-roundup-
epa-idUSKBN0E72IH20140527.

233 **Contrary to early beliefs:** Battaglin et al., "Glyphosate and Its Degradation Product AMPA."

233 **Research from South America:** Alejandra Paganelli et al., "Glyphosate-
Based Herbicides Produce Teratogenic Effects on Vertebrates by Impairing
Retinoic Acid Signaling," *Chemical Research in Toxicology* 23 (2010): 1586–95.

233 **One Canadian study:** T. E. Arbuckle, Z. Lin, and L. S. Mery, "An Exploratory Analysis of the Effect of Pesticide Exposure on the Risk of Spontaneous Abortion in an Ontario Farm Population," *Environmental Health Perspectives* 109, no. 8 (2001): 851–57.

233 **Other researchers have reported:** R. P. Rull, B. Ritz, and G. M. Shaw, "Neural Tube Defects and Maternal Residential Proximity to Agricultural Pesticide Applications," *Epidemiology* 15 (2004): s188.

233 **but the Swiss study was deemed:** European Network of Scientists for Social and Environmental Responsibility, "Lethal Effects of Genetically Modified Bt Toxin Confirmed on Young Ladybird Larvae," ScienceDaily, July 22, 2014, http://www.sciencedaily.com/releases/2012/02/120227111 158.htm and Angelika Hilbeck et al., "A Controversy Re-visited: Is the Coccinellid *Adalia bipunctata* Adversely Affected by Bt Toxins?" *Environmental Sciences Europe* 24 (2012), doi:10.1186/2190-4715-24-10.

234 **But a 2009 study found:** McLure, "Genetically Modified Food."

234 **As corn prices rose:** Tom Polansek, "Analysis: Facing Drought, U.S. Farmers Return to Crop Rotation," Reuters, January 18, 2013, http://www.reuters.com/article/2013/01/18/us-usa-drought-corn-idUSBRE90H0KR 20130118.

234 **Nearly half of US farmers surveyed:** Kent Fraser, "Glyphosate Resistant Weeds—Intensifying," Stratus Ag Research blog, January 25, 2013, http://www.stratusresearch.com/blog/glyphosate-resistant-weeds-intensifying.

234 **The problem was so bad:** Richard Davis, "RR Cotton Growers Can Receive Rebates for Multiple Herbicides," *NewsWatch*, February 2009, http://magissues.farmprogress.com/SCV/CV02Feb09/scv009.pdf.

234 **requiring biotech companies:** Dan Charles, "Insects Find Crack in Biotech Corn's Armor," *The Salt* (blog), National Pubic Radio, December 5, 2011, http://www.npr.org/blogs/thesalt/2011/12/05/143141300/insects-find -crack-in-biotech-corns-armor.

234 **a major ingredient in Agent Orange:** "Agent Orange: Background on Monsanto's Involvement," Monsanto, http://www.monsanto.com/news views/Pages/agent-orange-background-monsanto-involvement.aspx, accessed January 27, 2014.

235 **Indeed, chemical companies have invested:** Stephen O. Duke, "Why Have No New Herbicide Modes of Action Appeared in Recent Years?" *Pest Management Science* 68 (2012): 505–12.

235   **"I don't know about nitrogen-fixing corn":** Robert Fraley, interview by the author, January 13, 2009.

235   **"If you're waiting on private companies":** Bruce Babcock, interview by the author, April 26, 2012.

236   **"At full irrigation potential":** Qingwu Xue, interview by the author, January 25, 2013.

237   **by 2005 there was enough:** "Vitamin A Deficiency-Related Disorders (VADD)," Golden Rice Project, Golden Rice Humanitarian Board, http://www. goldenrice.org/Content3-Why/why1_vad.php, accessed February 15, 2013.

237   **Vitamin A deficiency has long been:** Ibid.

238   **Current vitamin A pill programs:** Tom Levitt, "Can GM-Free Biofortified Crops Succeed after Golden Rice Controversy?" *Ecologist*, December 12, 2011.

238   **The distribution of Golden Rice seeds:** "What Are the Levels of Carotenoid in the Donated Materials?" Golden Rice Project: Frequently Asked Questions, http://www.goldenrice.org/Content3-Why/why3_FAQ.php#Carotenoid_levels, accessed February 15, 2013.

238   **Farmers will get the seeds for free:** Ingo Potrykus, "The Golden Rice 'Tale,'" http://www.goldenrice.org/PDFs/The_GR_Tale.pdf, accessed February 15, 2013.

238   **GMO critics, however:** Ibid.

238   **Greenpeace has been aided in its cause:** Robin McKie, "After 30 Years, Is a GM Food Breakthrough Finally Here?" *Guardian*, February 2, 2013.

239   **"Golden Banana" for Africa:** Ibid.

239   **"crime against humanity":** Ingo Potrykus, "Lessons from the 'Humanitarian Golden Rice' Project: Regulation Prevents Development of Public Good Genetically Engineered Crop Products," *New Biotechnology* 27 (2010): 466–72.

239   **"To sequence the *Arabidopsis* genome":** Pamela Ronald, interview by the author, December 18, 2012.

240   **Some 50 million acres of rice:** "California Scientists Honored: Humanitarian Rice Research Is Making a Difference," RiceFarming.com, http:// www.ricefarming.com/home/issues/2013-01/California-Scientists-Honored .html, accessed January 28, 2014.

240   **Using genetic markers and DNA sequencing:** Ronald, interview, December 18, 2012; Pamela Ronald, *Tomorrow's Table* (blog), http://science

blogs.com/tomorrowstable; "New Flood-Tolerant Rice Offers Relief for World's Poorest Farmers," Ronald Laboratory, http://indica.ucdavis.edu/news/new-flood-tolerant-rice-offers-relief-for-worlds, accessed January 28, 2014; "Flood-Tolerant Rice Saves Farmers Livelihoods," International Rice Research Institute, http://irri.org/our-impact/increase-food-security/flood-tolerant-rice-saves-farmers-livelihoods, accessed January 28, 2014.

241   **With the new genetic tools:** Robert S. Ziegler and Adam Barclay, "The Relevance of Rice," *Rice* 1 (2008): 3–10.

241   **That's why many scientists:** James R. Ehleringer and Thure E. Cerling, "C_3 and C_4 Photosynthesis," in *Encyclopedia of Environmental Change*, vol. 2, *The Earth System: Biological and Ecological Dimensions of Global Environmental Change*, ed. Harold A. Mooney and J. G. Canadell (New York: Wiley, 2003), 186–90, http://www.ehleringer.net/Jim/Publications/271.pdf.

242   **"The bottom line is that":** Jane Langdale, e-mail message to the author, December 21, 2012.

242   **Experts estimate that the world:** Gilles van Kote, "Researchers Aim to Flick the High-Carbon Switch on Rice," *Guardian*, January 24, 2012.

242   **Global rice production in 2012:** "Better Weather Prompts Increase in Global Rice Production—UN Food Agency," UN News Centre, November 19, 2012, http://www.un.org/apps/news/story.asp?NewsID=43535&Cr=fao&Cr1=#.VAKVIWP4Kc0.

243   **Still, Langdale estimates:** Langdale, e-mail, December 21, 2012.

243   **Mark Lynas, a British journalist:** McKie, "After 30 Years."

243   **whose annual agricultural research budgets:** John King, Andrew Toole, and Keith Fuglie, "The Complementary Roles of the Public and Private Sectors in Agricultural Research and Development," Economic Brief 19 (USDA Economic Research Service, 2012), http://www.ers.usda.gov/media/913804/eb19.pdf.

244   **Yet federal funding is dwindling:** Tom Philpott, "How Your College Is Selling Out to Big Ag," *Mother Jones*, May 9, 2012.

Chapter 12. Organic Agriculture:
Feeding the Rich or Enriching the Poor?

247   **"The slow poisoning of the life":** Sir Albert Howard, *An Agricultural Testament* (New York: Oxford University Press, 1943), 220.

247   **The Grammy Awards have always been:** Bill Carter, "40 Million Watch

Grammys, with Whitney Houston in Viewers' Mind," *New York Times*, February 13, 2012.

248 **"Cultivate a better world":** "Back to the Start," Chipotle video, http://www.chipotle.com/en-US/fwi/videos/videos.aspx, March 12, 2013.

248 **the 29 million pounds of antibiotics:** Maryn McKenna, "News Break: FDA Estimates US Livestock Get 29 Million Pounds of Antibiotics per Year," *Wired*, December 9, 2010.

248 **"when practical":** "What Is Food with Integrity?" Chipotle, http://www.chipotle.com/en-US/fwi/fwi.aspx, April 24, 2014.

248 **But even such a nonbinding nod:** Maureen Morrison, "Chipotle Leaps Forward with 'Back to the Start,'" *Advertising Age*, November 26, 2012, http://adage.com/article/special-report-marketer-alist-2012/chipotle-leaps-forward-back-start/238415.

249 **Since 1990, organic food sales:** "Organic Agriculture: Overview," USDA Economic Research Service, last updated April 7, 2014, http://www.ers.usda.gov/topics/natural-resources-environment/organic-agriculture.aspx#.U1lL98fJ47A.

249 **By 2012 the US organic market:** Catherine Greene, "Growth Patterns in the U.S. Organic Industry," *Amber Waves*, October 24, 2013, http://www.ers.usda.gov/amber-waves/2013-october/growth-patterns-in-the-us-organic-industry.aspx#.VANL02P4Kc1.

249 **more than double Monsanto's sales:** Monsanto, "Annual Report 2013: Financial Highlights," http://www.monsanto.com/investors/pages/financial-highlights.aspx, accessed March 22, 2013.

250 **Urban gardens, farmer's markets:** "Urban Agriculture," *New York Times*, TimesTopics, http://topics.nytimes.com/top/reference/timestopics/subjects/a/agriculture/urban_agriculture, accessed January 30, 2014.

250 **Though studies vary widely:** Rachael L. Dettmann, "Organic Produce: Who's Eating It?" (paper presented at the American Agricultural Economics Association Annual Meeting, Orlando, FL, July 27–29, 2008).

250 **"We aren't going to feed":** Brian Halweil, "Can Organic Farming Feed Us All?" *World Watch* 19, no. 3 (May/June 2006), http://www.worldwatch.org/node/4060.

251 **In 1939, at age 20:** "Lady Eve Balfour," International Federation of Organic Agriculture Movements, http://www.ifoam.org/en/lady-eve-balfour, accessed April 2, 2013.

251 **"The ultimate goal of farming":** "Masanobu Fukuoka," International

Federation of Organic Agriculture Movements, http://www.ifoam.org/en/search?find=Masanobu, accessed April 2, 2013.

252 **"The research workers on most":** Sir Albert Howard, *The Soil and Health: A Study of Organic Agriculture* (Lexington: University Press of Kentucky, 2006), 246.

253 **"I was affected so profoundly":** J. I. Rodale, quoted on the book jacket of *An Agricultural Testament*, by Sir Albert Howard, special Rodale Press ed. (Emmaus, PA: Rodale Press, 1979).

255 **"When I became farm director":** Jeff Moyer, interview by the author, May 7, 2012.

257 **It means more fungi:** Darwin Anderson, "Organic Matter in Prairie Soils," *Crops and Soils*, January–February 2012.

257 **roughly 2,500 pentagrams of carbon:** Rattan Lal, "Crop Residues and Soil Carbon," http://www.fao.org/ag/ca/Carbon%20Offset%20Consultation/CARBONMEETING/3FULLPAPERSBYCONSULTATIONSPEAKERS/PAPERLAL.pdf, accessed March 12, 2014; Judith D. Schwartz, "Soil as Carbon Storehouse: New Weapon in Climate Fight?" *Yale Environment 360*, March 4, 2014, http://e360.yale.edu/feature/soil_as_carbon_storehouse_new_weapon_in_climate_fight/2744.

258 **The organic systems used 45 percent less energy:** Rodale Institute, "Fast Facts," in *The Farming Systems Trial: Celebrating 30 Years* (Emmaus, PA: Rodale Institute, 2011), http://66.147.244.123/~rodalein/wp-content/uploads/2012/12/FSTbookletFINAL.pdf; C. Peterson, L. Drinkwater, and P. Wagoner, *The Rodale Institute Farming Systems Trial: The First 15 Years* (Emmaus, PA: Rodale Institute, 1999).

258 **Other studies outside of Rodale:** Halweil, "Can Organic Farming Feed Us All?"

259 **harvesting 297 bushels per acre:** "Berks County Farmer Wins National Corn Grower Contest," News Release, County of Berks, Pennsylvania, http://www.co.berks.pa.us/Dept/AgCoord/Pages/FarmerWinsNatlCornGrowersContest.aspx, accessed April 12, 2013.

259 **"If the weather cooperates":** Moyer, interview, May 7, 2012.

260 **showing a 13 percent difference:** Verena Seufert, Navin Ramankutty, and Jonathan A. Foley, "Comparing the Yields of Organic and Conventional Agriculture," *Nature* 485 (2012): 229–32; Halweil, "Can Organic Farming Feed Us All?"

260 **with yields in Africa more than doubling:** *Organic Agriculture and Food Security in Africa* (United Nations Conference on Trade and Development,

United Nations Environment Programme, 2008), 16, http://unctad.org/en/Docs/ditcted200715_en.pdf; Halweil, "Can Organic Farming Feed Us All?"

261 **Global rice fields emit 13 percent:** Sarah Graham, "Rice Paddy Emissions Depend on Crops' Success," *Scientific American*, August 20, 2002; "Overview of Greenhouse Gases," Environmental Protection Agency, http://epa.gov/climatechange/ghgemissions/gases/ch4.html, accessed February 6, 2014.

261 **an uncanny 16 tons per hectare:** Norman Uphoff, "Development of the System of Rice Intensification (SRI) in Madagascar," Cornell University, http://sri.ciifad.cornell.edu/aboutsri/CIP_UPWARD_SRICase.pdf, accessed April 16, 2013.

262 **"They haven't invented this thing":** Amrik Singh, interview by the author, October 2008.

263 **The SRI plants were thicker:** Amrik Singh, "Performance of System of Rice Intensification and Conventional Rice Cultivation Methods under Punjab Conditions," http://ciifad.cornell.edu/sri/countries/india/InPunjab Rpt07.pdf, accessed April 17, 2013.

263 **In fact, Cornell researchers:** Norman Uphoff, interview by the author, May 3, 2013.

264 **"My father was really against it":** Makman Singh, interview by the author, October 2008.

264 **"Discussion of the system":** Beth Hoffman, "Can We Revolutionize Agriculture without Science?" *Forbes*, February 2, 2013.

264 **Critics at IRRI and elsewhere:** A. J. McDonald, P. R. Hobbs, and W. J. Riha, "Stubborn Facts: Still No Evidence That the System of Rice Intensification Outyields Best Management Practices (BMPs) Outside of Madagascar," *Field Crops Research* 108 (2008): 188–91.

265 **The method has been credited:** John Vidal, "India's Rice Revolution," *Guardian*, February 16, 2013.

266 **Uphoff credits the difference:** Russell J. Rodriguez et al., "Symbiotic Regulation of Plant Growth, Development and Reproduction," *Communicative & Integrative Biology* 2 (2009): 141–43.

266 **"The agricultural establishment said SRI":** Uphoff, interview, May 3, 2013.

266 **which is now practiced:** Vidal, "India's Rice Revolution."

267 **Like many other organic farmers:** Leontino Balbo Jr., interview by the author, April 2007; Native, http://www.nativealimentos.com.br/en.

267 **"The farmers in São Paulo":** Balbo, interview, April 2007.

268 **A recent study by FAO researchers:** Amir Kassam and Hugh Brammer, "Combining Sustainable Agricultural Production with Economic and Environmental Benefits," *Geographical Journal* 179 (2013): 11–18.

270 **"Right now we've got 4 percent":** Mark Smallwood, interview by the author, May 7, 2012.

Chapter 13. The Malawi Miracle

273 **"Let us pick any Malawi road":** Dr. Bingu wa Mutharika, acceptance speech, 2009 Drivers of Change Awards, October 29, 2009, Johannesburg, South Africa, http://www.southernafricatrust.org/speech_20091029.html, accessed July 8, 2013.

274 **Each year hunger kills more people:** "Hunger Kills More People Every Year Than AIDS, Malaria, & Tuberculosis Combined," World Food Programme, Hunger Statistics, http://www.wfp.org/hunger/stats, accessed March 5, 2014.

275 **"Some of the environmental lobbyists":** Gregg Easterbrook, "Forgotten Benefactor of Humanity," *Atlantic*, January 1997, http://www.the atlantic.com/magazine/archive/1997/01/forgotten-benefactor-of-humanity/ 306101.

275 **In his 70s Borlaug was pulled:** "SAA History in Brief," Sasakawa Africa Association, http://www.saa-safe.org/wwa/history.html, accessed June 28, 2013; "Where We Work: Ethiopia," Sasakawa Africa Association, http:// www.saa-safe.org/www/ethiopia.html, accessed June 28, 2013.

276 **Mbutu Sésé Seko, at the time "president for life":** Carole J. L. Collins, "Zaire/Democratic Republic of the Congo," *Foreign Policy in Focus*, July 1, 1997; Mary Anne Fitzgerald, "Obituary: Mobutu Sese Seko," *Independent*, September 9, 1997.

276 **Though such numbers are almost impossible:** Nick Dearden, "Africa's Wealth Is Being Devoured by Tyrants and Vultures," *Guardian*, July 21, 2012; "External Debt in Africa," Policy Brief 3, UN Office of the Special Adviser on Africa, October 2010, http://www.un.org/africa/osaa/ reports/2010_Debtbrief.pdf.

276 **Instead, the money just vanished:** James K. Boyce and Léonce Ndikumana, "Elites Loot Africa While Foreign Debt Mounts," *YaleGlobal*

Online, February 27, 2012, http://yaleglobal.yale.edu/content/elites-loot -africa-while-foreign-debt-mounts.

277 **while in sub-Saharan Africa:** Natasha Gilbert, "African Agriculture: Dirt Poor," *Nature*, March 28, 2012, http://www.nature.com/news/african -agriculture-dirt-poor-1.10311.

277 **Since 1960, average yields:** Felicia Wu and William Butz, *The Future of Genetically Modified Crops: Lessons from the Green Revolution* (Santa Monica, CA: Rand Corporation, 2004), 23, citing Sir Gordon Conway, "Biotechnology and Hunger" (remarks addressed to Senate about Science at the House of Lords, London, May 8, 2003).

277 **The continent now has an annual:** Joel K. Bourne Jr., "The End of Plenty," *National Geographic*, June 2009.

277 **If population, income, and urbanization continue:** Judith A. Chambers et al., "GM Agricultural Technologies for Africa: A State of Affairs," African Development Bank and IFPRI Report, 2014, 7, http://www.ifpri .org/sites/default/files/publications/pbs_afdb_report.pdf.

277 **"Sub-Saharan Africa is where the action is":** Pedro Sanchez, interview by the author, September 5, 2008.

278 **He spent $58 million:** Ibid.

279 **The *New York Times* wrote a story:** Celia W. Dugger, "Ending Famine, Simply by Ignoring the Experts," *New York Times*, December 2, 2007.

279 **Other African leaders jumped:** Zoé Druilhe and Jesús Barreiro Hurlé, *Fertilizer Subsidies in Sub-Saharan Africa*, ESA Working Paper 12-04 (Food and Agriculture Organization of the United Nations, Agricultural Development Economics Division, 2012), http://www.fao.org/docrep/016/ap077e/ap077e.pdf.

279 **Some economists have estimated:** John W. McArthur, "Fertilizing Growth: Estimating Agricultural Yields and Their Effects in Economic Development" (presentation to Oxford CSAE 2011, Conference on Economic Development in Africa, March 7, 2011).

280 **But big pushes cost money:** "Business Case: Millennium Village in Northern Ghana," United Kingdom Department for International Development, November 2011, http://www.cgdev.org/doc/2012/BusinessCase andSummary202483%282%29.pdf; Michael Clemons, "New Documents Reveal the Cost of Ending Poverty in a Millennium Village: At Least $12,000 per Household," *Global Development: Views from the Center* (blog), Center for Global Development, March 30, 2012, http://www.cgdev.org/

blog/new-documents-reveal-cost-"ending-poverty"-millennium-village
-least-12000-household.

281   **Cosmos Chimwara:** All conversations in the village of Mwandama and the
      "Justin" village—with Cosmos Chimwara, Mary Austin, Faison Tipoti,
      and Irene Maganga—took place with the author in October 2008.

283   **Thanks to enormous efforts:** "Where We Work: Africa: Malawi:
      Global Health," USAID, http://www.usaid.gov/malawi/global-health,
      accessed July 3, 2013; "Malawi," UNAIDS/WHO Epidemiological Fact
      Sheet, 2004 update, http://data.unaids.org/publications/fact-sheets01/
      malawi_en.pdf.

284   **Fewer than 10 percent of the households:** Generose Nziguheba et al.,
      "The African Green Revolution: Results from the Millennium Villages
      Project," *Advances in Agronomy* 109 (2010): 75–115.

284   **Many experts believe:** Olusequn Obasanjo, "How Africa Could Feed the
      World," CNN World, November 6, 2012, http://globalpublicsquare.blogs
      .cnn.com/2012/11/06/how-africa-could-feed-the-world.

285   **and the United Nations has projected:** "Population, Climate Change,
      and Sustainable Development in Malawi," Policy and Issue Brief, Popu-
      lation Action International, September 11, 2012, http://populationaction
      .org/policy-briefs/population-climate-change-and-sustainable-
      development-in-malawi.

286   **up to 75 percent of Malawi's calories:** "Malawi: Comprehensive Food
      Security and Vulnerability Analysis (CFSVA) and Nutrition Assessment,"
      World Food Programme and National Statistics Office of Malawi, Octo-
      ber 2012, http://documents.wfp.org/stellent/groups/public/documents/
      ena/wfp253658.pdf.

287   **"It was not an easy transition":** Rachel Bezner Kerr, interview by the
      author, May 9, 2013.

288   **"First, the fertilizer subsidy":** Zaharia Nkhonya, interview by the author,
      October 2008.

288   **"Before I joined the project":** The testimonies of SFHC farmers Ackim
      Mhone and Esnaly Ngwira were recorded by the author in October 2008.

290   **"Kudya!":** Alice Sumphi, interview by the author, October 2008.

291   **requiring large, back-to-back food lifts:** "Our Work: Comprehensive
      Food Security and Vulnerability Analysis (CFSVA)," World Food Pro-
      gramme, http://www.wfp.org/food-security/assessments/comprehensive-
      food-security-vulnerability-analysis, accessed June 25, 2013; "WFP Malawi
      2013 in Review: Toward a Food and Nutrition Secure and Resilient Malawi,"

World Food Programme, http://documents.wfp.org/stellent/groups/public/documents/communications/wfp266794.pdf, accessed June 14, 2014.

291 **Jeffrey Sachs was about the only one:** Jeffrey D. Sachs, "How Malawi Fed Its Own People," *New York Times*, April 19, 2012.

291 **The same year (2007) that Malawi reported:** Morten Jerven, "Agricultural Statistics: Who Benefits from Distortions?" World Economics, January 2103, http://www.worldeconomics.com/Papers/Agricultural%20Statistics%20Who%20benefits%20from%20distortions_96b727d6-fdf0-468e-b937-13680d2f7d6d.paper.

291 **"It was a great media blitz":** Thomas Jayne, interview by the author, June 6, 2013.

292 **Sachs has long refused to release:** "With Transparency Comes Trust" (editorial), *Nature* 485 (2012).

292 **In May of 2012, however:** Paul M. Pronyk et al., "The Effect of an Integrated Multisector Model for Achieving the Millennium Development Goals and Improving Child Survival in Rural Sub-Saharan Africa: A Non-randomised Controlled Assessment," *Lancet* 379 (2012): 2179–88.

292 **Unfortunately, their math was as bad as:** Gabriel Demombynes and Espen Beer Prydz, "The Millennium Villages Project Impacts on Child Mortality," *Development Impact* (blog), World Bank, May 10, 2012, http://blogs.worldbank.org/impactevaluations/the-millennium-villages-project-impacts-on-child-mortality.

293 **Only about half were still growing legumes:** Rachel Bezner Kerr et al., *Participatory, Agroecological and Gender-Sensitive Approaches to Improved Nutrition: A Case Study in Malawi* (FAO and WHO, 2013), http://www.fao.org/fileadmin/user_upload/agn/pdf/FAO-expert-meeting-submission-Bezner-Kerr-et-al-ver4-2_FAO_comments_doc.pdf.

293 **Such large plantations occupy 16 percent:** "Malawi," USAID Country Profile: Property Rights and Resource Governance, http://usaidlandtenure.net/sites/default/files/country-profiles/full-reports/USAID_Land_Tenure_Malawi_Profile.pdf, accessed October 2, 2014

294 **"If you wrote a letter to God":** Miguel Bosch, interview by the author, May 2013.

295 **"This could significantly reduce":** Gregory Meyers, interview by the author, May 2013.

296 **"This reduced shortfall is equivalent":** "Improving Access to Family Planning Can Promote Food Security in a Changing Climate—Study Summary: Modeling Climate Change, Food Security, and Population Growth,"

FS-12-71, MEASURE Evaluation PRH, March 2012, http://futuresgroup
.com/files/publications/Ethiopia_Brief_FINAL_FS-12-711_033012.pdf.

297 **"When people talk about adapting"**: Scott Moreland, quoted in "In Ethiopia, Combined Approaches May Be Key to Fighting Climate Change," MEASURE Evaluation PRH, January 17, 2012, http://www.cpc.unc.edu/measure/prh/in-ethiopia-combined-approaches-may-be-key-to-fighting-climate-change.

Chapter 14. The Grand *Desiderata*

299 **"We are not, however, to relax"**: T. R. Malthus, *An Essay on the Principle of Population, or, A View of Its Past and Present Effects on Human Happiness* (2nd ed., 1803), ed. Donald Winch, Cambridge Texts in the History of Political Thought (Cambridge: Cambridge University Press, 1992), 230.

301 **"that black and terrible demon"**: Philip Appleman, ed., introduction to *An Essay on the Principle of Population*, by Thomas Robert Malthus, Norton critical ed. (New York: Norton, 1976), xii.

301 **I later learned that his grave**: Patricia James, ed., "Biographical Sketches," in *The Travel Diaries of Thomas Robert Malthus* (London: Cambridge University Press, 1966), 14.

302 **Our ancestors domesticated wheat**: Manjit S. Kang and Surinder S. Banga, eds., *Combating Climate Change: An Agricultural Perspective* (Boca Raton, FL: CRC Press, 2013), 166; "Agriculture and Horticulture Turkey," Centre for Climate Adaptation, http://www.climateadaptation.eu/turkey/agriculture-and-horticulture, accessed March 14, 2014.

302 **"sixth great spasm of extinction"**: Edward O. Wilson, *The Diversity of Life* (Cambridge, MA: Belknap Press of Harvard University Press, 1992), 32.

302 **It's no coincidence:** World Wildlife Fund International, *Living Planet Report 2014: Species and Spaces, People and Places* (Gland, Switzerland: WWF International, 2014), 4.

303 **the "silver buckshot" approach:** Jonathan A. Foley, "Can We Feed the World and Sustain the Planet?" *Scientific American*, November 2011.

304 **Today, despite our far higher levels:** E. C. Oerke, "Centenary Review: Crop Losses to Pests," *Journal of Agricultural Science* 144 (2006): 31–43, doi:10.1017/S0021859605005708.

304 **"Providing food and nutrition":** "IonE-Led International Team Crafts Plan for Feeding the World While Protecting the Planet" (press release),

University of Minnesota, Institute of the Environment, October 12, 2011, http://environment.umn.edu/press-release/2011-news-archive.

304 **That's roughly the equivalent:** "Growing Greenhouse Gas Emissions Due to Meat Production," UN Environment Programme, UNEP Global Environmental Alert Service, October 2012, http://na.unep.net/geas/getUNEPPageWithArticleIDScript.php?article_id=92.

304 **So unhealthy is the average:** "Obesity Threatens to Cut U.S. Life Expectancy, New Analysis Suggests," *NIH News*, National Institutes of Health, March 16, 2005, http://www.nih.gov/news/pr/mar2005/nia-16.htm.

305 **If consumption trends continue:** Jane E. Allen, "Half of American Adults Headed for Diabetes by 2020, UnitedHealth Says," ABC News, November 25, 2010, http://abcnews.go.com/Health/Diabetes/diabetes-half-us-adults-risk-2020-unitedhealth-group/story?id=12238602.

305 **A recent study in the United Kingdom:** Ibid., citing P. Scarborough et al., "Modelling the Health Impact of Environmentally Sustainable Dietary Scenarios in the UK," *European Journal of Clinical Nutrition* 66 (2012): 710–15, doi:10.1038/ejcn.2012.34.

305 **More important, since both cattle:** "Eat Less Meat and Farm Efficiently to Tackle Climate Change, Scientists Say," ScienceDaily, June 19, 2012, http://www.sciencedaily.com/releases/2012/06/120619225934.htm.

305 **Of course, the most beneficial thing:** Malthus, *Essay*, 19; Appleman, introduction to *An Essay*.

305 **The United States and England:** Timothy Dyson, interview by the author, February 23, 2010.

306 **people were the "ultimate resource":** Julian L. Simon, *The Ultimate Resource* (Princeton, NJ: Princeton University Press, 1981).

306 **From 1995 to 2007:** J. Bongaarts, J. Cleland, J. W. Townsend, J. T. Bertrand, and M. Das Gupta, *Family Planning Programs for the 21st Century: Rationale and Design* (New York: Population Council, 2012), http://www.popcouncil.org/uploads/pdfs/2012_FPfor21stCentury.pdf.

307 **the "demographic dividend":** D. E. Bloom, D. Canning, and P. N. Malaney, *Demographic Change and Economic Growth in Asia*, CID Working Paper 15 (Center for International Development at Harvard University, 1999), http://www.hks.harvard.edu/var/ezp_site/storage/fckeditor/file/pdfs/centers-programs/centers/cid/publications/faculty/wp/015.pdf; D. E. Bloom et al., "Why Has China's Economy Taken Off Faster than India's?" June 2006,

http://isites.harvard.edu/fs/docs/icb.topic209735.files/Bloom_Canning _China_India.pdf.

308 **"That point, 1817":** Dyson, interview, February 23, 2010.

308 **"is the most important thing":** Tim Dyson, "On Democratic and Demographic Transitions," *Population and Development Review* 38 (2013): 83–102.

308 **Of the 25 countries left:** "2014 World Population Data Sheet," Population Reference Bureau, http://www.prb.org/Publications/Datasheets/2014/2014 -world-population-data-sheet.aspx, accessed March 4, 2014; "Total Fertility Rate, 1970 and 2013," Population Reference Bureau, http://www.prb .org/DataFinder/Topic/Rankings.aspx?ind=17, accessed March 4, 2014.

308 **A recent USAID report:** "Gender Equality and Women's Empowerment," USAID, http://www.usaid.gov/what-we-do/gender-equality-and-womens -empowerment, accessed March 4, 2014.

308 **"pink revolution":** Rena Singer (Landesa, Rural Development Institute), interview by the author, June 19, 2014.

309 **The Islamic Republic of Iran:** Elizabeth Leahy Madsen, "Iran: A Seemingly Unlikely Setting for World's Fastest Demographic Transition," *New Security Beat* (blog), Environmental Change and Security Program, Woodrow Wilson International Center for Scholars, January 11, 2012, http:// www.newsecuritybeat.org/2012/01/building-commitment-to-family-planning-iran.

309 **The strength of that system:** Joel K. Bourne Jr., "Iranian Cure for the Delta's Blues," *AARP Bulletin*, June 2010.

310 **This trend led Iran's former president:** Madsen, "Iran: A Seemingly Unlikely Setting."

310 **An increasing number of developing countries:** Ibid.

311 **The nation of Niger has:** "Niger: Overview," World Food Programme, http://www.wfp.org/countries/niger/overview, accessed March 4, 2014.

311 **Even if Niger's fertility rate is halved by 2050:** Bongaarts, *Family Planning Programs*, 20.

311 **In 2013, 10 million people:** "Sahel Food Insecurity Threatens 10 Million People in 2013," World Food Programme, News, February 14, 2013, http://www.wfp.org/content/sahel-food-insecurity-threatens-10-million -people-2013.

311 **It should surprise no one:** *Global Trends 2030: Alternate Worlds* (National Intelligence Council, 2012), http://www.dni.gov/nic/globaltrends.

311 **Strong family-planning programs:** Bongaarts, *Family Planning Programs*.

312 **"Family planning is just about the only thing"**: Dyson, interview, February 23, 2010.

312 **If all the world's 75 million:** Sophie Janaskie, "The Human Population Explosion," *Yale Scientific*, May 11, 2013, http://www.yalescientific.org/2013/05/the-human-population-explosion; Cory J. A. Bradshaw and Barry W. Brook, "Human Population Reduction Is Not a Quick Fix for Environmental Problems," *PNAS* 111 (2014): 16610–16615, doi: 10.1073/pnas.1410465111 accessed on 12/30/14.

312 **It has to go hand in hand:** Bradshaw and Brook, "Human Population Reduction."

313 **Yet despite more than two decades:** "NASA Scientists React to 400 ppm Carbon Milestone," National Aeronautics and Space Administration, Global Climate Change, News, http://climate.nasa.gov/400ppmquotes, accessed July 15, 2013.

313 **In fact, scientists believe:** Ibid.

314 **In March 2014 the International Panel:** IPCC Working Group II, *Climate Change 2014: Impacts, Adaptation, and Vulnerability*, AR5, Report by Chapters, http://www.ipcc.ch/report/ar5/wg2, accessed March 31, 2014.

316 **"just and equitable food and farming systems":** Center for Environmental Farming Systems, http://www.cefs.ncsu.edu, accessed September 4, 2014.

316 **The forester, conservationist, and farmer Aldo Leopold:** Aldo Leopold, *A Sand County Almanac* (New York: Oxford University Press, 1966), 262.

317 **"A thing is right when":** This is an exact quote from Leopold, *Sand County Almanac*, except the word "Earth" replaces the original phrase "biotic community."

SELECTED BIBLIOGRAPHY

Antoniou, M., M. E. M. Habib, C. V. Howard, R. C. Jennings, C. Leifert, R. O. Nodari, C. J. Robinson, and J. Fagan. "Teratogenic Effects of Glyphosate-Based Herbicides: Divergence of Regulatory Decisions from Scientific Evidence." *Journal of Environmental & Analytical Toxicology* S4 (2012).

Arbuckle, T. E., Z. Lin, and L. S. Mery. "An Exploratory Analysis of the Effect of Pesticide Exposure on the Risk of Spontaneous Abortion in an Ontario Farm Population." *Environmental Health Perspectives*, 109, no. 8 (2001): 851–57.

Arthus-Bertrand, Yann. *Earth from Above for Young Readers*. New York: Abrams, 2002.

Babcock, Bruce A. "Updated Assessment of the Drought's Impacts on Crop Prices and Biofuels Production." CARD Policy Brief 12-BP8. Center for Agricultural and Rural Development, Iowa State University, August 2012. http://www.card.iastate.edu/publications/synopsis.aspx?id=1169.

Barboza, David. "In China, Farming Fish in Toxic Waters." *New York Times*, December 15, 2007. http://www.nytimes.com/2007/12/15/world/asia/15fish.html?pagewanted=all.

Barringer, Felicity. "Empty Fields Fill Urban Basins and Farmers' Pockets." *New York Times*, October 23, 2011. http://www.nytimes.com/2011/10/24/science/earth/24water.html?pagewanted=all&_r=0.

Battaglin, W. A., M. T. Meyer, K. M. Kuivila, and J. E. Dietze. "Glyphosate and Its Degradation Product AMPA Occur Frequently and Widely in U.S. Soils, Surface Water, Groundwater, and Precipitation." *Journal of the American Water Resources Association (JAWRA)*, 50, no. 2 (2014): 275–90. doi:10.1111/jawr.12159.

Berry, Wendell. *The Unsettling of America: Culture and Agriculture*. San Francisco: Sierra Club Books, 1977.

Bezner Kerr, R., L. Shumba, L. Dakishoni, E. Lupafya, P. R. Berti, L. Classen, S. S. Snapp, and M. Katundu. *Participatory, Agroecological and Gender-*

Sensitive Approaches to Improved Nutrition: A Case Study in Malawi. FAO and WHO, 2013. http://www.fao.org/fileadmin/user_upload/agn/pdf/FAO -expert-meeting-submission-Bezner-Kerr-et-al-ver4-2_FAO_comments_ doc.pdf.

Bloom, David E., David Canning, Linlin Hu, Yuanli Liu, Ajay Mahal, and Winnie Yip. "Why Has China's Economy Taken Off Faster than India's?" June 2006. http://isites.harvard.edu/fs/docs/icb.topic209735.files/Bloom_Canning _China_India.pdf.

Bloom, David E., David Canning, and Pia N. Malaney. *Demographic Change and Economic Growth in Asia.* CID Working Paper 15. Center for International Development at Harvard University, 1999. http://www.hks.harvard .edu/var/ezp_site/storage/fckeditor/file/pdfs/centers-programs/centers/cid/ publications/faculty/wp/015.pdf.

Borlaug, Norman. "The Green Revolution, Peace, and Humanity" (Nobel lecture, December 11, 1970), http://www.nobelprize.org/nobel_prizes/peace/laureates/ 1970/borlaug-lecture.html.

Bovard, James. *Archer Daniels Midland: A Case Study in Corporate Welfare.* Cato Policy Analysis 241. Cato Institute, 1995. http://www.cato.org/pubs/pas/ pa-241.html.

Bowbrick, Peter. *A Critique of Professor Sen's Theory of Famines.* Oxford: Institute of Agricultural Economics, 1986.

Boyce, James K., and Léonce Ndikumana. "Elites Loot Africa While Foreign Debt Mounts." *YaleGlobal Online,* February 27, 2012. http://yaleglobal.yale .edu/content/elites-loot-africa-while-foreign-debt-mounts.

Breisinger, Clemons, Olivier Ecker, and Perrihan Al-Riffai. "Economics of the Arab Awakening: From Revolution to Transformation and Food Security." IFPRI Policy Brief 18. Washington, DC: International Food Policy Research Institute, 2011. http://www.ifpri.org/sites/default/files/publications/bp018 .pdf.

Brookes, Graham, and Peter Barfoot. "Key Environmental Impacts of Global Genetically Modified (GM) Crop Use 1996–2011." *GM Crops and Food* 4, no. 2 (2013): 109–19.

Brown, Lester. *Eco-Economy: Building an Economy for the Earth.* New York: Norton, 2001.

Brown, Lester. "Feeding Nine Billion." In *State of the World, 1999: A Worldwatch Institute Report on Progress toward a Sustainable Society.* Edited by Linda Starke. New York: Norton, 1999.

Brown, Lester. *Plan B 3.0: Mobilizing to Save Civilization*. Rev. ed. New York: Norton, 2008.

Brown, Lester. *Who Will Feed China?: Wake-up Call for a Small Planet*. New York: Norton, 1995.

Cappiello, Dina, and Matt Apuzzo. "The Secret Environmental Cost of US Ethanol Policy." AP News, November 12, 2013. http://bigstory.ap.org/article/secret-dirty-cost-obamas-green-power-push-1.

Carson, Rachel. *The Sea around Us*. Special ed. New York: Oxford University Press, 1991.

Carson, Rachel. *Silent Spring*. 40th anniv. ed. from Mariner Books. New York: Houghton Mifflin, 2002.

Cauvin, Henri E. "Between Famine and Politics, Zambians Starve." *New York Times*, August 30, 2002.

Ciezadlo, Annia. "Let Them Eat Bread: How Food Subsidies Prevent (and Provoke) Revolutions in the Middle East." *Foreign Affairs*, March 23, 2011. http://www.foreignaffairs.com/articles/67672/annia-ciezadlo/let-them-eat-bread.

Cleaver, Harry M., Jr. "The Contradictions of the Green Revolution." *American Economic Review* 62, no. 1/2 (1972): 177–86. http://www.jstor.org/stable/1821541.

Climate and Man: 1941 Yearbook of Agriculture. Washington, DC: US Government Printing Office, 1941.

Cohen, Joel E. "Seven Billion." *New York Times*, October 23, 2011. http://www.nytimes.com/2011/10/24/opinion/seven-billion.html?pagewanted=all.

Collins, Carole J. L. "Zaire/Democratic Republic of the Congo." *Foreign Policy in Focus*, July 1, 1997.

Coontz, Sydney H. *Productive Labour and Effective Demand: Including a Critique of Keynesian Economics*. London: Routledge & Kegan Paul, 1965.

Cullather, Nick. *The Hungry World: America's Cold War Battle against Poverty in Asia*. Cambridge, MA: Harvard University Press, 2010.

Dearden, Nick. "Africa's Wealth Is Being Devoured by Tyrants and Vultures." *Guardian*, July 21, 2012.

Demombynes, Gabriel, and Espen Beer Prydz. "The Millennium Villages Project Impacts on Child Mortality." *Development Impact* (blog), World Bank, May 10, 2012. http://blogs.worldbank.org/impactevaluations/the-millennium-villages-project-impacts-on-child-mortality.

De Silva, Sena S., and F. Brian Davy, eds. *Success Stories in Asian Aquaculture*. New York: Springer, 2009.

Devereux, Stephen. "Famine in the Twentieth Century." Institute of Development Studies, Working Paper 105, January 1, 2000. http://www.ids.ac.uk/publication/famine-in-the-twentieth-century.

Dixon, John, and Aidan Gulliver. *Farming Systems and Poverty: Improving Farmers' Livelihoods in a Changing World*. Edited by Malcolm Hall. Rome: FAO and World Bank, 2001. http://www.fao.org/3/a-ac349e.pdf.

Donthi, Praveen. "Cancer Express." *Hindustan Times*, January 16, 2010. http://www.hindustantimes.com/News-Feed/India/Cancer-Express/Article1-498286.aspx.

Dugger, Celia W. "Ending Famine, Simply by Ignoring the Experts." *New York Times*, December 2, 2007.

Dyson, Tim. "On Democratic and Demographic Transitions." *Population and Development Review* 38 (2012): 83–102.

Dyson, Tim. *Population and Food: Global Trends and Future Prospects*. London: Routledge, 1996.

Dyson, Tim, and Arup Maharatna. "Excess Mortality during the Bengal Famine: A Re-evaluation." *Indian Economic and Social History Review* 28, no. 3 (1991): 281–97.

Easterbrook, Gregg. "Forgotten Benefactor of Humanity." *Atlantic*, January 1997. http://www.theatlantic.com/magazine/archive/1997/01/forgotten-benefactor-of-humanity/306101.

Elwell, Frank W. *A Commentary on Malthus' 1798 Essay on Population as Social Theory*. Lewiston, NY: Mellen, 2001.

Environmental Working Group. "MTBE in Drinking Water." October 22, 2013. http://www.ewg.org/research/mtbe-drinking-water.

Famine Inquiry Commission. *Report on Bengal*. Government of India, 1945.

Firth, Peta. "Leaving a Bad Taste." *Scientific American*, May 1999.

Flynn, Laurie, and Michael Gillard. "Pro-GM Food Scientist 'Threatened' Editor." *Guardian*, October 31, 1999.

Foley, Jonathan A. "Can We Feed the World and Sustain the Planet?" *Scientific American*, November 2011, 60–65. http://www.scientificamerican.com/article/can-we-feed-the-world.

Food and Agriculture Organization of the United Nations. "How to Feed the World in 2050." High Level Expert Forum, Food and Agriculture Organization of the United Nations, Rome, October 12–13, 2009. http://www.fao.org/wsfs/forum2050/wsfs-forum/en.

Food and Agriculture Organization of the United Nations. *Save and Grow: A Policymaker's Guide to Sustainable Intensification of Smallholder Crop Production.* Rome: FAO, 2011. http://www.fao.org/docrep/014/i2215e/i2215e.pdf.

Food and Agriculture Organization of the United Nations. *The State of Fisheries and Aquaculture, 2012.* http://www.fao.org/docrep/016/i2727e/i2727e.pdf.

Food and Agriculture Organization of the United Nations. *The State of Fisheries and Aquaculture, 2014.* http://www.fao.org/3/a-i3720e.pdf.

Freedom from Famine: The Norman Borlaug Story (film). [Dayton, OH]: Mathile Institute for the Advancement of Human Nutrition, 2009.

Friedman, Thomas L. *Hot, Flat, & Crowded: Why the World Needs a Green Revolution—And How We Can Renew Our Global Future,* release 2.0. New York and London: Penguin, 2009.

Fukuoka, Masanobu. *The One-Straw Revolution: An Introduction to Natural Farming.* New York: NYRB Classics, 2009. First published 1975.

Funk, Chris C., and Molly E. Brown. "Declining Global Per Capita Agricultural Production and Warming Oceans Threaten Food Security." *Food Security* 1 (2009): 271–89. doi:10.1007/s12571-009-0026-y.

Gilbert, Natasha. "African Agriculture: Dirt Poor." *Nature,* March 28, 2012. http://www.nature.com/news/african-agriculture-dirt-poor-1.10311.

Gillis, Justin. "A Warming Planet Struggles to Feed Itself." *New York Times,* June 4, 2011.

Gleick, Peter H. "Dam It, Don't Dam It, Undam It: America's Hydropower Future." *Huffington Post,* August 6, 2012. http://www.huffingtonpost.com/peter-h-gleick/americas-hydropower-future_b_1749182.html.

Gleick, Peter H. "Zombie Water Projects (Just When You Thought They Were Really Dead . . ." *Forbes.com,* December 7, 2011. http://www.forbes.com/sites/petergleick/2011/12/07/zombie-water-projects-just-when-you-thought-they-were-really-dead.

Goswami, Omkar. "The Bengal Famine of 1943: Re-examining the Data." *Indian Economic and Social History Review* 27 (1990): 445–63. doi:10.1177/001946469002700403.

Haddok, Eitan. "Biofuels Land Grab: Guatemala's Farmers Lose Plots and Prosperity to 'Energy Independence.'" *Scientific American,* January 13, 2012. http://www.scientificamerican.com/article.cfm?id=biofuels-land-grab-guatemala.

Halweil, Brian. "Can Organic Farming Feed Us All?" *World Watch* 19, no. 3 (May/June 2006). http://www.worldwatch.org/node/4060.

Hansen, Chad. "Daoism." In *The Stanford Encyclopedia of Philosophy*. Fall 2013 ed. Edited by Edward N. Zalta. http://plato.stanford.edu/archives/fall2013/entries/daoism.

Hargrove, Tom, and W. Ronnie Coffman. "Breeding History." *Rice Today* 5, no. 4 (2010): 34–38.

Hettel, Gene. "Luck Is the Residue of Design." IRRI Pioneer Interviews. *Rice Today* 5, no. 4 (2010): 10.

Hilbeck, Angelika, Joanna M. McMillan, Matthias Meier, Anna Humbel, Juanita Schläpfer-Miller, and Miluse Trtikova. "A Controversy Re-visited: Is the Coccinellid *Adalia bipunctata* Adversely Affected by Bt Toxins?" *Environmental Sciences Europe* 24 (2012): art. 10. doi:10.1186/2190-4715-24-10.

Hine, Rachel, and Jules Pretty. *Organic Agriculture and Food Security in Africa*. United Nations Conference on Trade and Development, United Nations Environment Programme, 2008. http://unctad.org/en/Docs/ditcted200715_en.pdf.

Hoff, Benjamin. *The Tao of Pooh*. New York: Penguin, 1983.

Howard, Sir Albert. *An Agricultural Testament*. Special Rodale Press ed. Emmaus, PA: Rodale Press, 1979. First published 1943 by Oxford University Press.

Howard, Sir Albert. *The Soil and Health: A Study of Organic Agriculture*. Lexington: University Press of Kentucky, 2006. First published 1947 by Devon-Adair Company.

Huggins, David R., and John P. Reganold. "No-Till: The Quiet Revolution." *Scientific American*, July 2008, 70–77.

Intergovernmental Panel on Climate Change. *Climate Change 2013: The Physical Science Basis*. Cambridge: Cambridge University Press, 2013. http://www.climatechange2013.org.

Intergovernmental Panel on Climate Change. *Climate Change 2014: Impacts, Adaptation, and Vulnerability*. Cambridge: Cambridge University Press, 2014. http://ipcc-wg2.gov/AR5.

Intergovernmental Panel on Climate Change. *Climate Change 2014: Synthesis Report*. Cambridge: Cambridge University Press, 2014. http://www.ipcc.ch/pdf/assessment-report/ar5/syr/AR5_SYR_FINAL_SPM.pdf.

Jackson, Wes. *Becoming Native to This Place*. Berkeley, CA: Counterpoint, 1996.

Jackson, Wes. *New Roots for Agriculture*. Lincoln: University of Nebraska Press, 1980.

James, Patricia. *Population Malthus: His Life and Times*. London: Routledge & Kegan Paul, 1979.

Janaskie, Sophie. "The Human Population Explosion." *Yale Scientific*, May 11, 2013. http://www.yalescientific.org/2013/05/the-human-population-explosion.

Jerven, Morten. "Agricultural Statistics: Who Benefits from Distortions?" World Economics, January 2013. http://www.worldeconomics.com/Papers/Agricul tural%20Statistics%20Who%20benefits%20from%20distortions_96b727d6 -fdf0-468e-b937-13680d2f7d6d.paper.

Kang, Manjit S., and Surinder S. Banga, eds. *Combating Climate Change: An Agricultural Perspective.* Boca Raton, FL: CRC Press, 2013.

Kassam, Amir, and Hugh Brammer. "Combining Sustainable Agricultural Pro-duction with Economic and Environmental Benefits." *Geographical Journal* 179 (2013): 11–18.

Katz, Alan, and Peter Robison. "Morgan Stanley Bet the Farm in Ukraine before Taxpayers' Bailout." Bloomberg, October 4, 2011. http://www.bloomberg .com/news/2011-10-04/morgan-stanley-bet-the-farm-in-ukraine-before-fed -bailout-by-u-s-taxpayer.html.

King, John, Andrew Toole, and Keith Fuglie. "The Complementary Roles of the Public and Private Sectors in Agricultural Research and Development." Economic Brief 19. USDA Economic Research Service, 2012. http://www .ers.usda.gov/media/913804/eb19.pdf.

King, John Kerry. "Rice Politics," *Foreign Affairs* 31 (1953): 453–60. http://www .jstor.org/stable/20030978.

Knight, Sir Henry. *Food Administration in India, 1939–47.* Stanford, CA: Stan-ford University Press, 1954.

Kovarik, Bill. "Henry Ford, Charles Kettering and the Fuel of the Future." *Auto-motive History Review*, no. 32 (Spring 1998): 7–27. http://www.environmen talhistory.org/billkovarik/about-bk/research/henry-ford-charles-kettering -and-the-fuel-of-the-future.

Lagi, Marco, Yavni Bar-Yam, Karla Z. Bertrand, and Yaneer Bar-Yam. "The Food Crises: A Quantitative Model of Food Prices Including Speculators and Ethanol Conversion." New England Complex Systems Institute, September 21, 2011. http://necsi.edu/research/social/food_prices.pdf.

Lal, Rattan. "Crop Residues and Soil Carbon." http://www.fao.org/ag/ca/Carbon %20Offset%20Consultation/CARBONMEETING/3FULLPAPERSBY CONSULTATIONSPEAKERS/PAPERLAL.pdf.

Latham, Michael E. *The Right Kind of Revolution: Modernization, Development, and U.S. Foreign Policy from the Cold War to the Present.* Ithaca, NY: Cornell University Press, 2011.

Leopold, Aldo. *A Sand County Almanac*. Reprint ed. New York: Ballantine, 1986.

Li, Wenhua, ed. *Agro-ecological Farming Systems in China*. Man and the Biosphere Series 26. Paris: UNESCO, 2001.

Loftas, Tony, ed. *Dimensions of Need: An Atlas of Food and Agriculture*. Rome: Food and Agriculture Organization of the United Nations, 1995. http://www.fao.org/docrep/U8480E/U8480E00.htm.

Lovei, Magda. *Phasing Out Lead from Gasoline: Worldwide Experience and Policy Implications*. World Bank Technical Paper 397. Washington, DC: World Bank, 1998. http://siteresources.worldbank.org/INTURBANTRANSPORT/Resources/b09phasing.pdf.

Madsen, Elizabeth Leahy. "Iran: A Seemingly Unlikely Setting for World's Fastest Demographic Transition." *New Security Beat* (blog), Environmental Change and Security Program, Woodrow Wilson International Center for Scholars, January 11, 2012. http://www.newsecuritybeat.org/2012/01/building-commitment-to-family-planning-iran.

Malthus, T. R. *An Essay on the Principle of Population, or A View of Its Past and Present Effects on Human Happiness*. Edited by Donald Winch. Cambridge Texts in the History of Political Thought. Cambridge: Cambridge University Press, 1992.

Malthus, T. R. *An Essay on the Principle of Population: Text, Sources and Background Criticism*. Edited by Philip Appleman. New York: Norton, 1976.

Malthus, T. R. *The Travel Diaries of Thomas Robert Malthus*. Edited by Patricia James. London: Cambridge University Press, 1966.

Marsh, George Perkins. *Man and Nature*. Edited by David Lowenthal. Cambridge, MA: Belknap Press of the Harvard University Press, 1965. Reprint, Seattle: University of Washington Press, 2003. First published 1864.

Maternal and Child Nutrition: Executive Summary of The Lancet *Maternal and Child Nutrition Series*. Lancet, [2013]. http://download.thelancet.com/flat contentassets/pdfs/nutrition-eng.pdf.

McKelvey, John J., Jr. *J. George Harrar 1906–1982: A Biographical Memoir*. Biographical Memoirs 57. Washington, DC: National Academy of Sciences, 1987.

McLure, Jason. "Genetically Modified Food." *CQ Researcher*, August 31, 2012, 717–39.

Mitchell, Donald. *A Note on Rising Food Prices*. Policy Research Working Paper 4682. World Bank Development Prospects Group, July 2008. https://openknowledge.worldbank.org/bitstream/handle/10986/6820/WP4682.pdf.

Molden, David, ed. *Water for Food, Water for Life: A Comprehensive Assessment of Water Management in Agriculture.* London: Earthscan, 2007. http://www.iwmi.cgiar.org/assessment/Publications/books.htm.

Montgomery, David R. *Dirt: The Erosion of Civilizations.* Berkeley: University of California Press, 2007.

Montgomery, David R. "Soil Erosion and Agricultural Sustainability." *Proceedings of the National Academy of Sciences USA* 104, no. 33 (2007): 13268–72. doi:10.1073/pnas.0611508104.

Moreland, Scott, and Ellen Smith. "Modeling Climate Change, Food Security and Population," MEASURE Evaluation PRH, February 2012. http://www.cpc.unc.edu/measure/publications/sr-12-69.

Moroz, Serhiy. "Rural Households in Ukraine: Current State and Tendencies." *Economics of Agriculture* 60 (2013): 565–84. http://purl.umn.edu/158256.

Morris, Christopher W., ed. *Amartya Sen.* New York: Cambridge University Press, 2010.

National Aeronautics and Space Administration. "NASA Scientists React to 400 ppm Carbon Milestone." Global Climate Change, News. http://climate.nasa.gov/400ppmquotes.

National Intelligence Council. *Global Trends 2030: Alternative Worlds.* December 2012. http://www.dni.gov/nic/globaltrends.

National Research Council. *Renewable Fuel Standard: Potential Economic and Environmental Effects of U.S. Biofuel Policy.* Washington, DC: National Academies Press, 2011. http://www.nap.edu/openbook.php?record_id=13105&page=383.

Nomura. *The Coming Surge in Food Prices.* Global Economics and Strategy. Nomura, 2010. http://www.nomura.com/europe/resources/pdf/080910.pdf.

Nziguheba, G., C. A. Palm, T. Berhe, G. Denning, A. Dicko, O. Diouf, W. Diru, et al. "The African Green Revolution: Results from the Millennium Villages Project." *Advances in Agronomy* 109 (2010): 75–115.

Ó Gráda, Cormac. *Famine: A Short History.* Princeton, NJ: Princeton University Press, 2009. http://press.princeton.edu/chapters/s8857.pdf.

Ó Gráda, Cormac. "The Ripple That Drowns? Twentieth-Century Famines in China and India as Economic History," *Economic History Review,* 61 (2008): 5–37.

Ó Gráda, Cormac. *"Sufficiency and Sufficiency and Sufficiency": Revisiting the Bengal Famine of 1943–44.* UCD Centre for Economic Research Working Paper

WP10/21. Dublin: UCD School of Economics, University College Dublin, 2010.

Organisation for Economic Co-operation and Development. "OECD Environmental Outlook to 2050: The Consequences of Inaction." 2012. http://www.oecd.org/environment/oecdenvironmentaloutlookto2050theconsequences ofinaction.htm.

Ortiz, R., D. Mowbray, C. Dowswell, and S. Rajaram. "Dedication: Norman E. Borlaug, the Humanitarian Plant Scientist Who Changed the World." *Plant Breeding Reviews* 28 (2007).

Padmanabhan, S. Y. "The Great Bengal Famine." *Annual Review of Phytopathology* 11 (1973): 11–24. doi:10.1146/annurev.py.11.090173.000303.

Paganelli, A., V. Gnazzo, H. Acosta, S. L. López, and A. E. Carrasco. "Glyphosate-Based Herbicides Produce Teratogenic Effects on Vertebrates by Impairing Retinoic Acid Signaling." *Chemical Research in Toxicology* 23 (2010): 1586–95.

Peterson, C., L. Drinkwater, and P. Wagoner. *The Rodale Institute Farming Systems Trial: The First 15 Years.* Emmaus, PA: Rodale Institute, 1999.

Philpott, Tom. "How Your College Is Selling Out to Big Ag." *Mother Jones*, May 9, 2012.

Potrykus, Ingo. "The Golden Rice 'Tale.'" http://www.goldenrice.org/PDFs/The_GR_Tale.pdf.

Potrykus, Ingo. "Lessons from the 'Humanitarian Golden Rice' Project: Regulation Prevents Development of Public Good Genetically Engineered Crop Products." *New Biotechology* 27 (2010): 466–72.

Pronyk, Paul M., et al. "The Effect of an Integrated Multisector Model for Achieving the Millennium Development Goals and Improving Child Survival in Rural Sub-Saharan Africa: A Non-randomised Controlled Assessment." *Lancet* 379 (2012): 2179–88.

Reisner, Marc. *Cadillac Desert: The American West and Its Disappearing Water.* Rev. ed. New York: Penguin, 1993.

Reisner, Marc. "The Fight for Reclamation." *High Country News*, March 20, 1995. http://www.hcn.org/issues/31/874.

Richerson, R. Boyd, and R. L. Bettinger. "Was Agriculture Impossible during the Pleistocene but Mandatory during the Holocene? A Climate Change Hypothesis." *American Antiquity* 66 (2001): 387–411. http://www.sscnet.ucla.edu/anthro/faculty/boyd/AgOrigins.pdf.

Rodale Institute. *The Farming Systems Trial: Celebrating 30 Years.* Emmaus, PA:

Rodale Institute, 2011. http://66.147.244.123/~rodalein/wp-content/up loads/2012/12/FSTbookletFINAL.pdf.

Sachs, Jeffrey D. "The End of Poverty." Twelfth annual Zuckerman Lecture, British Association for the Advancement of Science, London, 2005.

Sachs, Jeffrey D. *The End of Poverty: Economic Possibilities for Our Time.* New York: Penguin, 2005.

Sachs, Jeffrey D. "How Malawi Fed Its Own People." *New York Times*, April 19, 2012.

Sainath, P. "Farm Suicides: A 12-Year Saga." *Hindu*, January 25, 2010.

Sainath, P. "In 16 Years, Farm Suicides Cross a Quarter Million." *Hindu*, October 29, 2011.

Sainath, P. "Some States Fight the Trend but . . ." *Hindu*, December 5, 2011. http://www.thehindu.com/opinion/columns/sainath.

Schiermeier, Quirin. "Ecologists Fear Antarctic Krill Crisis." *Nature* 467, no. 15 (2010). doi:10.1038/467015a.

Sen, Amartya. *Poverty and Famines: An Essay on Entitlement and Deprivation.* New York: Oxford University Press, 1981.

Sen, Amartya, and Jean Drèze. *The Amartya Sen and Jean Drèze Omnibus: Comprising Poverty and Famines, Hunger and Public Action, India: Economic Development and Social Opportunity.* New Delhi: Oxford University Press, 1999 (10th impression, 2006).

Seufert, Verena, Navin Ramankutty, and Jonathan A. Foley. "Comparing the Yields of Organic and Conventional Agriculture." *Nature* 485 (2012): 229–32.

Shiva, Vendana. *Manifestos on the Future of Food and Seed.* New York: South End Press, 2007.

Shiva, Vendana. *The Violence of the Green Revolution.* London: Zed Books, 1992.

Simon, Julian L. *The Ultimate Resource.* Princeton, NJ: Princeton University Press, 1981.

Stanway, David. "China Says Over 3 Million Hectares of Land Too Polluted to Farm." Reuters, December 30, 2013. http://www.reuters.com/article/2013/12/30/china-environment-farmland-idUSL3N0K90OY20131230.

Swaminathan, M. S. "From Green to an Ever-Green Revolution." In *Science and Sustainable Food Security: Selected Papers of M. S. Swaminathan.* Singapore: World Scientific, 2009. doi:10.1142/9789814282116_0001.

Swaminathan, M. S. "Memorial Address." Norman Borlaug memorial service,

Texas A&M University, October 6, 2009. http://borlaug.tamu.edu/files/2010/10/Borlaug-Memorial-Swaminathan-Remarks.pdf.

Tauger, Mark B. "Entitlement, Shortage, and the 1943 Bengal Famine: Another Look." *Journal of Peasant Studies* 31 (2003): 60–63.

Tilman, D., J. Fargione, B. Wolff, C. D'Antonio, A. Dobson, R. Howarth, D. Schindler, W. H. Schlesinger, D. Simberloff, and D. Swackhamer. "Forecasting Agriculturally Driven Global Environmental Change." *Science* 292 (2001): 281. doi:10.1126/science.1057544.

Tiwana, N. S., N. Jerath, S. S. Ladhar, G. Singh, R. Paul, D. K. Dua, and H. K. Parwana. *State of Environment: Punjab-2007.* Chandigarh, India: Punjab State Council for Science & Technology, 2007. http://www.npr.org/documents/2009/apr/punjab_report.pdf.

Turral, Hugh, Jacob Burke, and Jean-Marc Faurès. *Climate Change, Water and Food Security.* FAO Water Reports 36. Rome: Food and Agriculture Organization of the United Nations, 2011. http://www.fao.org/docrep/014/i2096e/i2096e.pdf.

United Nations Educational, Scientific and Cultural Organization. *Managing Water under Uncertainty and Risk.* United Nations World Water Development Report 4. Paris: UNESCO, 2012. http://unesdoc.unesco.org/images/0021/002156/215644e.pdf.

United Nations Environment Programme. "Growing Greenhouse Gas Emissions Due to Meat Production." UNEP Global Environmental Alert Service, October 2012. http://na.unep.net/geas/getUNEPPageWithArticleIDScript.php?article_id=92.

United Nations Environment Programme. *International Assessment of Agricultural Knowledge, Science and Technology for Development.* Nairobi: UNEP, 2008.

United Nations Environment Programme. "Oil Palm Plantations: Threat and Opportunities for Tropical Ecosystems." UNEP Global Environmental Alert Service (GEAS), December 2011. http://www.unep.org/pdf/Dec_11_Palm_Plantations.pdf.

United Nations Environment Programme. *Organic Agriculture—A Step towards the Green Economy in the Eastern Europe, Caucasus and Central Asia Region—Case Studies from Armenia, Moldova and Ukraine.* Green Economy. Geneva: UNEP, 2011. http://www.unep.ch/etb/publications/Green%20Economy/Organic%20Agriculture%20-%20EECCA%20study%20on%20Green%20Economy/UNEP%20Organic_EECCA%20country%20study.pdf.

United Nations Population Division. "Seven Billion and Growing: The Role of

Population Policy in Achieving Sustainablility." Technical Paper no. 2011/3. New York: UN Department of Economic and Social Affairs, 2011. http://www.un.org/esa/population/publications/technicalpapers/TP2011-3_SevenBillionandGrowing.pdf.

United Nations Population Division. *World Population Prospects: The 2015 Revision*. New York: UN Department of Economic and Social Affairs, Population Division, 2015. http://esa.un.org/unpd/wpp/.

United Nations Population Division. *World Population to 2300*. New York: UN Department of Economic and Social Affairs, Population Division, 2004. http://www.un.org/esa/population/publications/longrange2/WorldPop 2300final.pdf.

US Congress, Government Accountability Office. *Seafood Safety: FDA Needs to Improve Oversight of Imported Seafood and Better Leverage Limited Resources.* Report to Congressional Requesters, GAO-11-286. US Government Accountability Office, 2011. http://www.gao.gov/assets/320/317734.pdf.

US Congress, Office of Technology Assessment. "Gasohol: A Technical Memorandum." September 1979. http://books.google.com/books/about/Gasohol .html?id=iNRTAAAAMAAJ.

US Department of Agriculture Foreign Agricultural Service. "Ukraine: Grain Production Prospects for 2008–2009." Commodity Intelligence Report, May 15, 2008. http://www.pecad.fas.usda.gov/highlights/2008/05/ukr_15may 2008/ukr_15may2008.htm.

US Environmental Protection Agency. *Achieving Clean Air and Clean Water: Report of the Blue Ribbon Panel on Oxygenates in Gasoline.* EPA-420-R-99-021. Environmental Protection Agency, 1999. http://www.epa.gov/otaq/consumer/fuels/oxypanel/r99021.pdf.

Vidal, John. "India's Rice Revolution." *Guardian*, February 16, 2013.

Wang, Michael, May Hu, and Hong Ho. "Life-Cycle Energy and Greenhouse Gas Emission Impacts of Different Corn Ethanol Plant Types." *Environmental Research Letters* 2, no. 2 (2007): 12. doi:10.1088/1748-9326/2/2/024001.

Warnemuende, Elizabeth A., Judodine P. Patterson, Douglas R. Smith, and Chi-hua Huang. "Effects of Tilling No-Till Soil on Losses of Atrazine and Glyphosate to Runoff Water under Variable Intensity Simulated Rainfall." *Soil and Tillage Research* 95 (2007): 19–26. doi:10.1016/j.still.2006.09.001.

Warren, Rachel. "The Role of Interactions in a World Implementing Adaptation and Mitigation Solutions to Climate Change." *Philosophical Transactions* of the Royal Society. A: Mathematical, Physical & Engineering Sciences 369 (2011): 217–241. doi:10.1098/rsta.2010.0271.

Wilson, E. O. Biophilia. Reprint ed. Cambridge, MA: Harvard University Press, 1986.

Winch, Donald. Malthus. New York: Oxford University Press, 1987.

"With Transparency Comes Trust" (Editorial). Nature 485 (2012).

Wolff, Gary, and Peter H. Gleick. "The Soft Path for Water." In The World's Water 2002–2003, chap. 1. Washington, DC: Island Press, 2002. http://pacinst.org/publication/the-worlds-water-2002-2003.

World Bank. Competitive Agriculture or State Control: Ukraine's Response to the Global Food Crisis. World Bank, Europe and Central Asia Region Sustainable Development Unit, 2008. http://siteresources.worldbank.org/INT UKRAINE/Resources/WorldFoodCrisisandRoleofUkraine.pdf.

World Bank. World Development Report 2010: Development and Change. Washington, DC: World Bank, 2010.

World Bank and OECD. Achieving Ukraine's Agricultural Potential: Stimulating Agricultural Growth and Improving Rural Life. Washington, DC: OECD and World Bank, 2004. http://www.oecd.org/tad/agricultural-policies/34031855.pdf.

World Food Programme and National Statistics Office of Malawi. "Comprehensive Food Security and Vulnerability Analysis (CFSVA) and Nutrition Assessment." World Food Programme, October 2012. http://documents.wfp.org/stellent/groups/public/documents/ena/wfp253658.pdf.

Wrigley, E. A., and R. S. Schofield. The Population History of England, 1541–1871: A Reconstruction. Cambridge: Cambridge University Press, 1989.

Wu, Felicia, and William Butz. The Future of Genetically Modified Crops: Lessons from the Green Revolution. Santa Monica, CA: Rand Corporation, 2004. http://www.rand.org/pubs/monographs/MG161.html.

Zerbe, Noah. "Feeding the Famine? American Food Aid and the GMO Debate in Southern Africa." Food Policy 29 (2004): 593–608.

Zhang, Jialin. "China's Slow-Motion Land Reform: Tentative Steps and Halting Progress." Policy Review, February 1, 2010.

Ziegler, Robert S., and Adam Barclay. "The Relevance of Rice." Rice 1 (2008): 3–10.

ILLUSTRATION CREDITS

Introduction: John Stanmeyer / VII Photo Agency

Chapter 1: Courtesy of the British Museum

Chapter 2: William Vandivert / Life Picture Collection / Getty Images

Chapter 3: Courtesy International Maize and Wheat Improvement Center (CIMMYT)

Chapter 4: Prashanth Vishwanathan / Bloomberg / Getty Images

Chapter 5: John Stanmeyer / VII Photo Agency

Chapter 6: Robert Clark / Institute

Chapter 7: Ed Kashi / VII Photo Agency

Chapter 8: Brian J. Skerry / National Geographic Creative

Chapter 9: Michael Nicholson / Corbis

Chapter 10: Dick Durrance II / National Geographic Creative

Chapters 11–14: By the author, with help from Yann Arthus-Bertrand, who generously allowed a photograph from his *Earth from the Air* exhibit to appear in the image of Bath Abbey

INDEX

Note: *Italic* page numbers refer to illustrations.